Sustainable Mobility

Edited by Bernardo Llamas, Marcelo F. Ortega Romero and Eugenia Sillero

Published in London, United Kingdom

IntechOpen

Supporting open minds since 2005

Sustainable Mobility
http://dx.doi.org/10.5772/intechopen.82169
Edited by Bernardo Llamas, Marcelo F. Ortega Romero and Eugenia Sillero

Contributors
Enrique Espi, Iñigo Ribas, Carlos Diaz, Oscar Sastron, Carlos Repáraz, Ewa Stawiarska, Hüseyin Turan Arat, Bahattin Tanç, Nevzat Özaslan, Gonzalo Fernandez-Sanchez, Sinan Erdogan, Eugenia Sillero, Paloma Martinez, Carla Garcia, Elena Martínez, Cynthia V. V. González-López, Francisco G. Acién-Fernández, Jose M. Fernández-Sevilla, Natalia Jawiarczyk, Francisco García Cuadra, Pedro Cotera, Manuel Arias

First published in London, United Kingdom, 2020 by IntechOpen
IntechOpen is the global imprint of INTECHOPEN LIMITED, registered in England and Wales, registration number: 11086078, 7th floor, 10 Lower Thames Street, London, EC3R 6AF, United Kingdom
Printed in Croatia

British Library Cataloguing-in-Publication Data
A catalogue record for this book is available from the British Library

Additional hard and PDF copies can be obtained from orders@intechopen.com

Sustainable Mobility
Edited by Bernardo Llamas, Marcelo F. Ortega Romero and Eugenia Sillero
p. cm.
Print ISBN 978-1-78984-562-4
Online ISBN 978-1-78984-563-1
eBook (PDF) ISBN 978-1-78984-620-1

An electronic version of this book is freely available, thanks to the support of libraries working with Knowledge Unlatched. KU is a collaborative initiative designed to make high quality books Open Access for the public good. More information about the initiative and links to the Open Access version can be found at www.knowledgeunlatched.org

For EU product safety concerns:
IN TECH d.o.o., Prolaz Marije Krucifikse Kozulić 3, 51000 Rijeka, Croatia, info@intechopen.com or visit our website at intechopen.com.

Meet the editors

Bernardo Llamas is a Lecturer at the Universidad Politécnica de Madrid, UPM. He has worked with the private sector in R&D departments and in the search for technologies to combat climate change and alternative fuel production. Thus, he collaborated on projects dealing with carbon dioxide capture for storage and carbon dioxide sequestration using microalgae systems and the production of bio-methane for the automobile sector. Currently he is involved in several projects to store energy in a feasible way. As well as his research and teaching on project management, he is also coordinating several actions at the Project Management Laboratory at the Higher School of Mining and Energy Engineering.

Eugenia Sillero runs the association that promotes the use of natural gas, biomethane and hydrogen in road and sea transport, Gasnam. She has a Master degree in Mine engineering specializing in energy and fuel from the Universidad Politécnica de Madrid and a PhD in communication by CEU San Pablo University. She has 25 years' work experience in both the public and private sectors always linked to sustainability. Her professional career began in the automotive components industry and continues in the Ministry of Transport where she held various positions of responsibility until joining Gasnam.

Marcelo F. Ortega Romero is an Associate Professor of Process Engineering and Statistical Data Analysis in theSchool of Mining and Energy of the Universidad Politécnica de Madrid from 2011. He has published 25 research papers in journals with an impact index in the Journal Citation Report and has attended more than 30 international congresses. His specialization and participation in research projects has been related to environmental pollution, CO2 geologic storage, characterization and production of biocarburants and analysis and optimization of energetic systems.

Contents

Preface

Greenhouse gas emissions continue to grow year after year and, despite efforts, the goal of limiting the concentration of CO_2 in the atmosphere to no more than 450 ppm (Paris Protocol): the CO_2 concentration recorded in Mauna Loa now exceeds 410 ppm, with an annual increase of more than 2 ppm. Thus, in the fight against climate change, the reduction of emissions must correspond to all industrial sectors. Notably, mobility and the use of fossil fuels pose a challenge towards sustainability in globalization and the transport of people and goods.

Mobility and its implications have a very important weight in the framework of sustainable development. The continuous growth that this sector has been experiencing over the last years and its foreseeable increase make the challenge of achieving sustainable transport a strategic priority at local, national, European and global levels. According to International Energy Agency, in 2017, 28.89 % of the total world primary energy was consumed by the transportation sector with an annual growth of around 4% since 1990. With regard to CO_2 emissions, according to the same agency (IEA), the transport sector was responsible for 24.5% in 2017. The CO_2 emission associated with the transport sector has been increasing by an average of 200 million tons per year since 1990.

The challenge to achieve sustainable mobility is great and this involves a great effort by everyone. For this, it is necessary to know the available technologies that make transportation more sustainable around the world day by day. We have to remember that not all technologies are applicable in all cities or towns and that the most important thing is efficient management of available resources.

The package of European measures and regulations drives the objective of obtaining new fuels that are based on renewable sources (biofuels) or the transformation of waste into fuels and the use of alternative fuels such as natural gas and hydrogen. But, the challenge in reaching a significant share in the use of these alternative fuels is in achieving logistics and distribution that reaches the end user. The European Directive 2014/94/EC encourages the establishment of a distribution network for these fuels considered alternative.

The objective of this book is to compile the examples and advances that the different industrial sectors have made: from the automotive sector to that of more sustainable conventional fuels, taking in to account the passenger's transport and including an example from some emerging countries. Recent lines of research are also included, for example on obtaining fuels from plastic waste is developed and the degradation of organic matter to obtain renewable natural gas. The first chapters will give the reader an overview about the use of alternative and sustainable fuels for transport, while more specific chapters have been grouped into biofuels, alternative fuels and electricity.

In this context, innovation seems fundamental to overcome the marked challenges of achieving neutrality in greenhouse gas emissions. And it will be creativity and capacity for sustainable leadership that will allow humanity to achieve this goal.

Therefore, we hope that this book contributes to increasing the creative capacity of the readers, finding results achieved in the various research projects and, serving as an incentive for you, the future reader, to take an active part in this challenge of humanity: to reduce the GHG emissions.

This book joins the publishing trajectory of Prof. Bernardo Llamas, who has contributed to several books already edited by IntechOpen, with diverse themes such as Greenhouse, CO_2 capture and storage, creativity and innovative project. But this time, other experts collaborate as editors: Prof. Marcelo F. Ortega Romero and Dr Eugenia Sillero as experts on alternative fuels (hydrogen and green natural gas) support this new initiative. All editors are taking part in a new R&D project to develop green natural gas: LIFE SMART AgroMobility. The aim of this project is transforming organic matter into biofuel. This biofuel (biogas) will be upgraded into biomethane to be used in light vehicles.

We hope that this new book will be a stimulus to promote creative solutions for sustainable mobility and we encourage you to be part of the solution in reducing CO_2 emissions.

Bernardo Llamas and Marcelo F. Ortega Romero
Universidad Politécnica de Madrid,
ETSI Minas y Energía,
Madrid, Spain

Eugenia Sillero
GASNAM,
Madrid, Spain

Section 1

Introduction

Chapter 1

The Pathway to Sustainable Transport

Pedro Cotera and Manuel Arias

Abstract

In 2015 the 17 United Nations (UN) Sustainable Development Goals (SDGs) by 195 countries were developed and agreed. The target is to end poverty, protect the planet, and ensure prosperity for all. Sustainable mobility and transport can be considered one of the main topics within this ambitious plan, considering its transversal influence in many of the 17 goals. In a world driven by global trends like climate change, local emissions, population growth, urbanization, emerging markets, digitalization, etc., a quick proactiveness to shift the mobility and transport to a sustainable way is mandatory. Taking all these drivers into a more practical level directly linked to mobility and transport, we can summarize them into four: congestion, local emissions, climate change, and energy security. There are many technologies and services to work on these areas. We consider three pillars as the umbrella to reach sustainable transport: energy efficiency, alternative fuels, and smart transport. In this chapter we will develop these three main pillars about what we can do already today without waiting some decades (probably will be too late in that case) but also looking into the future to give a neutral and realistic view. Why not begin already with the rolling fleet? If we train the drivers of a transport fleet reducing, for example, a 7% of fuel consumption, the carbon footprint will be reduced with 7% as well. Why wait 20 years until a new technology is developed? There are many opportunities in alternative fuels as well, already with competitive costs. Probably not all of them will be the solution everywhere, but they cannot be rejected. Some other alternatives like biomethane are a global solution for a circular economy with a huge potential to reduce local emissions and climate change and solve problems due to urbanization growth. We will try to explain why biofuels together with electrification are needed and why only electrification is not enough. Smart transport will be also covered speaking about which possibilities are available to make more efficient and safety transport and mobility, like bigger trucks or busses, or the introduction of advance driver assistance system (ADAS) in a new scenario.

Keywords: sustainability, mobility, transport, autonomous driving, truck, bus, car, efficiency, alternative fuel, congestion, emission, climate change, biofuel, biomethane, biogas, electrification

1. Introduction

There is an urgent need to brake the climate change [1]. The planet temperature is increasing quickly, with clear consequences like since 2001 we have had 18 of the 19 warmest years on record [2] (**Figure 1**).

Humanity has probably one of the most important challenges has had ever, and decisions cannot be postponed. Determination and global collaboration are needed to achieve the goal, which does not have a single solution. Each area of the society and economy has the responsibility and the opportunity to collaborate. **Figure 2** shows the CO_2 emissions by sector [3].

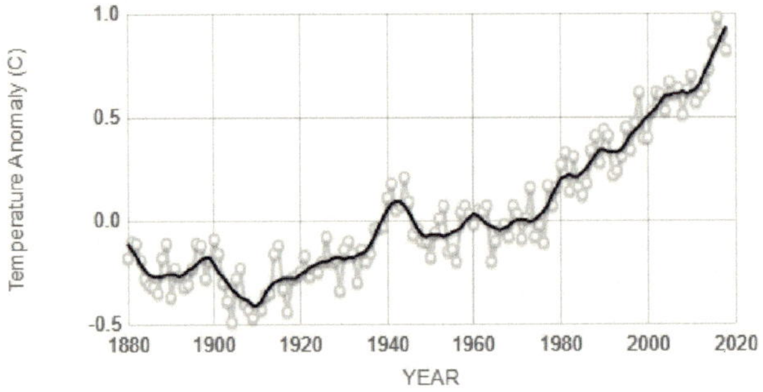

Figure 1.
Temperature evolution on Earth. Source: climate.nasa.gov.

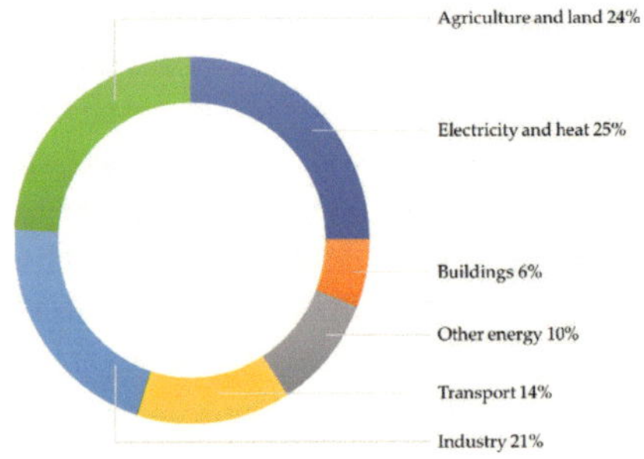

Figure 2.
CO_2 emissions by sector.

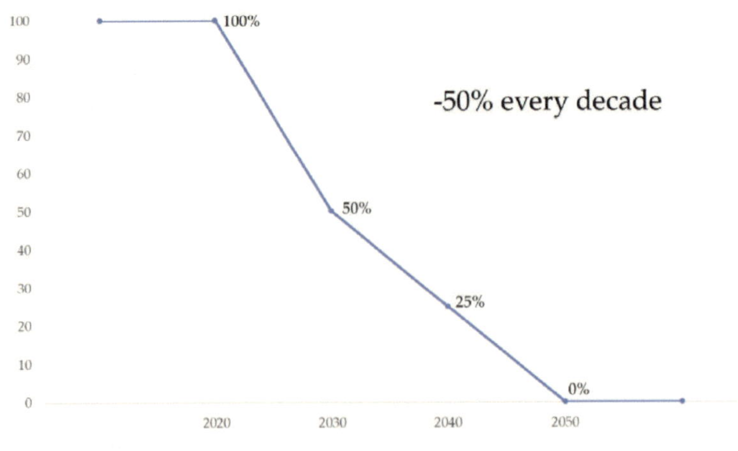

Figure 3.
Carbon law.

In this chapter we will focus on the solutions for transport sector, which represents the 14% of the total emissions, but no sector or country can walk alone to reach such ambitious target. That's the main reason of the Paris Agreement [1] and 17 UN SDGs. Due to their importance, we want to mention this agreement and targets, although it is not the goal of this chapter. We invite the readers to get more information since they will be the base for the coming years' decisions at global industry and economy.

Johan Rockström and some other prominent scientists have published in 2017 in "science" what they call "carbon law" [4]. They propose a strategy with 75% chance to reach the target of Paris Agreement [1] to keep below 2°C the increasement of the average planet temperature compared to pre-industrial temperature. Basically, they say that it must, and can, reduce 50% CO_2 emissions every decade until 2050 (**Figure 3**).

A realistic roadmap of real actions is what we proposed on the following article. Some of them are ready to begin right now, and some others must be developed during the following years. It is needed to work parallel both roads; we cannot only focus today or future but both. Otherwise we will fail.

It is honest and not a demagogy that both authors would like to leave a better world for our children and the coming generations. A strong commitment from everyone is needed, and we have an opportunity to change our daily actions.

We will try to make an easy read article, easy for everyone in or out of the scientist area. The target is so important that this kind of information must be spread everywhere and to everyone.

2. Drivers for mobility and transport

Although we mainly speak about climate change considering the emergency of the consequences it has, in making a deep analysis, we can find more drivers. Solutions will be present during the rest of the article, but it is appropriate to stop and analyze the scenario and problems we are facing to.

There are four main drivers for sustainable mobility and transport:

2.1 Congestion

According to the UN [5], by 2030 it is projected that 68% of the population will live in urban areas, compared to 30% on 1950 and 58% on 2019. These figures mean a lot of good opportunities but also an increase on congestion and pollution. Frequently it is only considering the alternative fuel as a solution for the mobility problems, but we can full the cities of, for example, electric cars and bikes, and that means pollution problem will be solved, but congestion will still remain a problem in the city. At cities like Tokyo with 37 million inhabitants [4], congestion is a real problem.

2.2 Pollution

At the same time, pollution causes millions of premature deaths annually due to household and mobility gases. These are dramatic figures that must be considered by the governments in the action plans of the cities. Some cities in Europe are already making new mobility regulations, restricted access areas for cars, etc., to reduce pollution problem and the consequences it has on the population and economy. Congestion is also very related to these new movements at city governments.

2.3 Energy security

There are financial and economical worldwide strategies among many countries. Energy security is one of the keys nowadays within that scenario. There are lot of movements in the industry and governments looking for being independent of whom could be a competitor on this economic and technological race. Every movement is positive if it helps the common target of achieving a sustainable world.

2.4 Climate change

All drivers are important, but the most urgent and important is climate change. The consequences can be so radical if we do not change anything that actions are mandatory from daily details up to international level.

It is important that everyone works in solution for these drivers in our daily small actions, but for long term and global solution, it is also crucial that an ecosystem is created between the main actors involved into the decisions that can drive the shift to a sustainable transport. From our point of view, it is crucial that this ecosystem of close collaboration must be formed by governments, energetic companies, transport companies, vehicle manufacturer, and end-customer (industry, logistic center, operator, etc.). All must be involved in the decisions as an active part.

3. Three pillars for sustainable transport

Nowadays, there are many areas, concepts, theories, classifications, lists, etc. explained in companies and reports like possible solutions to the problems we are facing to. Digitalization and electrification are frequently identified like the "coolest" and are frequently described in articles. From our point of view, there are three more global areas that we proposed to be the pillars for this pathway for a sustainable transport.

These pillars are alternative fuels, energy efficiency, and smart transport. We will try to explain each one in the following points.

4. Alternative fuels

We consider three factors to classify an alternative fuel: climate impact, availability, and cost. These factors are the key to evaluate if a technology is mature enough to expand or if at least can be used in an area of the planet or if it must be discarded. For example, in the transport sector, the operative cost is crucial to keep alive the company; we cannot forget it, so cost is important as well. But it cannot be the decision point, since everyone knows that for a new sustainable technology, there is an implementation curve with a higher price at the beginning. The solution is a close collaboration within the ecosystem before mentioned. There must be opened partnerships and dialogs involving all actors to facilitate the implementation of these new sustainable technologies, with enough availability and affordable price. The final customer, private company or only one actor of that ecosystem, cannot support the total cost.

On the other hand, alternative fuels are commonly identified as electrification. Of course, electrification will be crucial on the change to a sustainable transport, but it alone will not be enough. There is a race to reduce CO_2 emissions as quickest as possible. Otherwise, if we wait too much to act, nature inertia will be even more difficult to stop.

But there is good news. There are already available several alternative fuels that can be used today reducing emissions considerably and fulfilling the three criteria exposed. We will describe the main options available today and during the coming decades.

Before going into details, it is basic to be clear and give a realistic and complete picture of the alternative fuels. Alternative fuels have something that we call "life cycle assessment" (LCA), which means to consider the emissions coming from the whole process of the fuel life until recycling, including production (also batteries and minerals needed), transport, etc. Other ways to do it are only considering a part of the process: well-to-wheel (WtW), well-to-tank (WtT), or tank-to-wheel (TtW). There must be several ways to measure and analyze a process; it is positive depending on the use we give to the fuel. But there must be a final target and a complete cycle must be considered. There are fuels working in a combustion process, for example, biogas, that in the whole cycle can have much lower greenhouse effect than a battery electric vehicle (BEV). As much as realistic we consider the analysis, it will be much helpful for the target which is to reduce climate effect. Otherwise, we are lying ourselves.

4.1 Biofuels

We need to slope down quickly the emissions. To reach that in a big scale, the only and immediate alternative option we have are biofuels used in combustion engines.

Within biofuels we can find many options. Not all of them are available in every country, but taking the opportunity of what is reachable in the area is the correct action to move forward. We will try to mention the most relevant options available in the market.

Biogas, also called as renewable gas or biomethane when applied at mobility, is probably the most extended and accessible everywhere. The reason is because the resource is also everywhere: it is produced mainly from farm and city wastes. Speaking before about the considerable increase of inhabitants per city during the coming decades, we are also speaking about a potential problem on how to handle that huge amount of waste, or we can speak about a tremendous potential of biogas production for energy at industry, buildings, and mobility.

Already today, countries like Sweden have 94% of gas used for mobility with origin bio [6]. This is a perfect example of circular economy. What could be a problem is transformed into an opportunity. Swedish society is very well aware that there is a strong dialog between the actors in the ecosystem. Legislators, industry, customers, etc. have a common target (**Figure 4**).

At Sweden, 64% of gas produced ends up into upgrading mainly for mobility [7] (cars, trucks, busses, and engines).

There is another important fact regarding biogas: it is the base for the coming hydrogen economy. High percentage of industry and infrastructure will be valid for hydrogen technology. We will not go into details about hydrogen as biofuel since it needs a book itself, but we wanted to emphasize the importance of its link with biogas.

Hydrotreated vegetable oil (HVO) is also an interesting alternative fuel, produced by renewable raw materials. It can be used in diesel engines partially blended or 100% pure. Depending on the source, it can reduce 50–60% or even up to 90% CO_2 and can also reduce local emissions like NOx. HVO has better long-term storage stability and cold climate properties. It is already used frequently blended at commercial diesel fuel to reach bio quota.

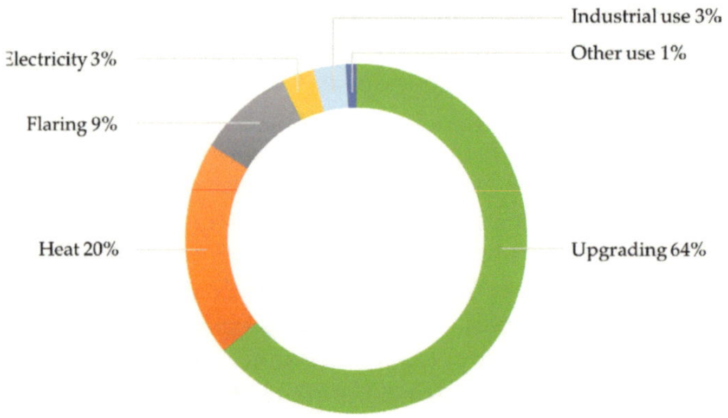

Figure 4.
End use of biogas production at Sweden.

HVO seems to be a perfect fuel from a technical property point of view, but it has very limited availability because of market reasons. Production is very limited worldwide, there is a high demand of HVO to be used and blended, so the price is high. On the other hand, availability at public stations is very limited in some countries like Sweden, and in many others, there is no even local provider.

We have tried several times to bring it to Spain for commercial operations, but it failed due to business case. Hopefully it will be a positive scenario in the market for the coming years to use in more extended way these kinds of biofuels.

Bioethanol is also a very important biofuel. It is produced by fermentation of sugars coming from feedstocks like wheat or corn. It can work blended with petrol or can be used blended with 95% additive like at Scania engines.

Bioethanol does not produce SO_2 or NOx and can reduce CO_2 up to 90% depending on the production cycle and source.

As in the HVO case, infrastructure is not developed, and in the case of being interested, the infrastructure must be customized and developed for the customer.

In the case of biofuels like biogas, HVO and bioethanol, a collaboration between ecosystem actors to develop appropriate production availability, infrastructure, tax regulation, and clear information to end-customer is urgent.

An argument frequently used against biofuels is that if they increase the production, food price will also increase. In case they come from waste, it is not true or in case if they come from crops either. It is a polemic issue, and there are many literatures speaking about positive or negative impacts. We can easily conclude based on **Figure 5** (source: Food and Agriculture Organization of the United Nations, FAO [8]) that vegetable oil price has no relation to prices of sugar, meat, cereals, or dairy products.

4.2 Electrification

There are many information about electrification mobility. We will not add new technical information but will try to make a sustainable approach and reflections.

Electrification is one of the most important keys in the pathway to a sustainable mobility. But it must be clear that nature will not wait until a mature technology is developed totally for passenger cars and, more difficult, for trucks and busses. Bear in mind the current autonomies and prices for the technology and infrastructure for a "simple" passenger car and then compare it to the requirement demands when transporting 40 or 60 tons at 4000 kilometers or when transporting 55 people in a bus. Even more, those transports cannot only be possible but also profitable.

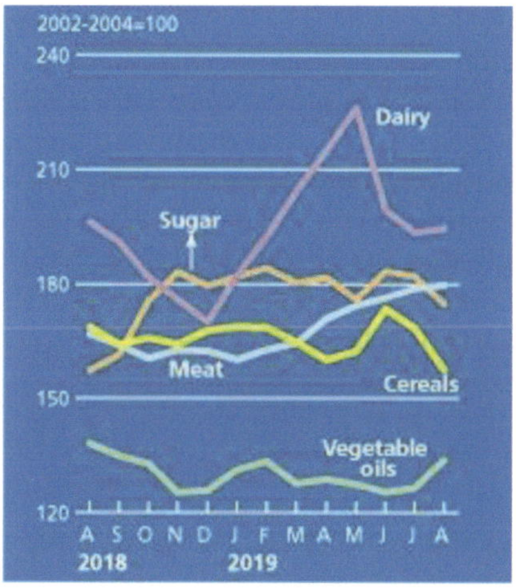

Figure 5.
FAO food commodity price indices.

This is the main reason to justify that society and scientist community must forget the war between technologies, keeping only a war against climate change where must be place for technologies that allows to reduce CO_2 footprint. That's the reason of a very high importance of biofuels for the coming decades. Carbon law becomes crucial when speaking about the pointless question of "which is the only valid and saving technology?" The answer is none, but several are needed quickly in a first step. Nobody knows the alternative fuel scenario at 2050, but surely will not be one single energy and will have to coexist several.

Still speaking about electrification, there are good electrification and bad electrification. We must consider the LCA of the vehicle, including mineral extraction and recycling or second life of the batteries. Then, the technology is valid and positive. The same case with electricity source must be considered. If we use the energy coming from a coal plant, for example, we can speak about zero emissions locally, but we are lying to ourselves speaking about climate change.

On the other hand, electrification is commonly linked to batteries, but there are many other ways to electrify a route. For example, there are some electrified roads where some dozens of trucks are running in Sweden, Germany, and soon in Italy. Probably it seems strange to go back to pantograph technology, but the results until now show that infrastructure is cheaper and results are impressive. The target is not

Figure 6.
Real e-highway in Germany.

the installation in every road but only in some main transport routes. The impact is still under study, but preliminary figures say that only a small percentage of main route trucks working with this technology would save millions of CO_2 tons per year.

The truck can run on diesel or biofuels when is out on electrified road, and during the time connected can also charge a battery to run the vehicle on electric out of the road for around 10 km (**Figure 6**).

We would like to leave a very positive message for electrification but also a responsibility strong message that it is not so easy to think electricity is equal to clean electrification. No, we must go deep in the details and get a real clean electrification.

5. Energy efficiency

This is the simplest pillar to understand and reach: keep the most efficient level of the technology you already have, the way of working, the operations you make, and your daily actions. Simple. For example, if you use the car with a colleague to the office every day instead of going alone, you are doubling the efficient emissions and cost. Another example, if you can reduce the fuel consumption of your vehicle 10%, you will also reduce your cost in the same percentage, but also emissions will be cut the same amount.

This is a very important idea to be considered in our life and businesses. It does not matter what technology to use, but an efficient way of doing things will impact into a more sustainable operation. Another way to explain as we mentioned before, a traffic jump road is the same mobility problem with BEV or petrol cars. The sustainable mobility does not depend only on the energy but also on the efficient way in which we move people and goods.

Another example is the trend of making transport to city center shopping areas during the night with big trucks using alternative fuels. The centers of big cities have tough congestion problems during the day, and an easy way to solve them is to deliver the goods during the night.

Connectivity and digitalization are the key to reach the top efficient level in mobility and transport. On 2012, transport market began to change thanks to connectivity. Until that moment, no transport company had information about what was happening on the truck. The company only knew the fuel consumption and incidents afterwards, but no information have the detail information enough to take decisions and make the company more efficient, more sustainable, and more profitable. For example, basic information like if the driver brakes too strong or driving with idling. When connectivity appeared, driver environment changed completely. Today it is possible to train the drivers not from a theoretical way, but doing it customized for each driver, each person, help him to improve all areas of their daily job.

There are already driver training programs for long term. Our experience is that when training a driver, probably next day he will drive in the same way he was used to before training. But thanks to a coaching program we can follow the driver and way of driving and figure and help him to improve his skills as much as possible.

There is a huge potential on this area, with a tremendous quick and easy implementation opportunity.

6. Smart transport

We consider smart transport in every new advanced technology that helps transport and mobility to be more efficient and sustainable, for example, bigger trucks where we increase the Tn/km-liter or m3/km-liter.

New technologies like platooning or autonomous driving are helping already to make a more suitable transport.

Autonomous driving is already implemented in real operation test at closed environments like mines and probably soon in other scenarios like close logistic centers, airports, bus depot, etc. This will be a reality in a very short term. Those closed places, which because of security or healthy reasons are more positive to have an autonomous driving truck or bus, will be more sustainable (**Figures 7** and **8**).

It is not the moment to make a deep analysis of autonomous driving, but we clearly see this technology with a predominant role in the sustainable transport for coming decades.

Platooning is another way of transport to improve sustainability. It is not a technology itself, but it takes advantage of every technology available. The main target is to reduce the distance between three of four vehicles to reduce air resistance. As lower distance, better aerodynamic, so better fuel consumption and much lower emissions. Companies like Scania are making this kind of transport since 2014 between its factories at Sweden and Netherlands (**Figure 9**).

This is possible because of the philosophy "right here, right now." It means that it is not needed to wait decades for a new technology; we can already begin with current technology. Frequent platooning is related to autonomous driving and probably in the future will be. But today we can already begin to do it with the current

Figure 7.
Autonomous driving Scania truck working in a closed area.

Figure 8.
Truck working in an underground mine. An example where autonomous driving can help.

Figure 9.
Platooning truck transport.

technology available at the truck, like advance control cruise. As technology evolves, the improvement will be even higher and emissions will be lower. The most important message is that we can already begin today and improve as the technology does.

7. Conclusion

Sustainable mobility and transport are crucial. We urgently need help from every area of the society to fight against the climate change. Human beings are not aware about the problem they are facing to, and we have a short time to react.

But we would like to leave a positive message. There are a lot of opportunities right here that can lead to the solution, and it is totally in our hands. As we have tried to explain, there are many solutions, technology, services, etc. that can already begin today to reduce emissions and even more in the coming decades.

We have the tools, and it is upon us to reach the target.

Author details

Pedro Cotera[1] and Manuel Arias[2*]

1 Scania, Zaragoza, Spain

2 Scania, Madrid, Spain

*Address all correspondence to: manuel.arias@scania.com

References

[1] Paris Agreement. 2015. Available from: https://unfccc.int/sites/default/files/english_paris_agreement.pdf [Accessed: 01 September 2019]

[2] NASA. 2019. Available from: climate.nasa.gov [Accessed: 01 September 2019]

[3] IPCC. In: Core Writing Team; Pachauri RK, Meyer LA, editors. Climate Change 2014: Synthesis Report. Contribution of Working Groups I, II and III to the Fifth Assessment Report of the Intergovernmental Panel on Climate Change. Geneva, Switzerland: IPCC; 2014. 151 pp

[4] Rockström J, Gaffney O, Rogelj J, et al. A roadmap for rapid decarbonization. Science. 2017;**355**(6331)

[5] United Nations, Department of Economic and Social Affairs, Population Division. World Urbanization Prospects: The 2018 Revision, Online Edition. File 21: Annual Percentage of Population at Mid-Year Residing in Urban Areas by region, subregion and country, 1950-2050. 2018

[6] Energigas Sverige. 2019. Available from: https://www.energigas.se/om-oss/nyheter-och-press/nyheter/94-procent-biogas-i-den-svenska-fordonsgasen/ [Accessed: 20 August 2019]

[7] Energigas Sverige. Proposal for National Biogas Strategy 2.0. 2018. Available from: https://www.energigas.se/library/2303/national-biogas-strategy-2_0.pdf [Accessed: 01 September 2019]

[8] Food and Agriculture Organization of the United Nations (FAO). 2019. Available from: http://www.fao.org/worldfoodsituation/foodpricesindex/en/ [Accessed: 05 September 2019]

Chapter 2

Evolution towards a Sustainable Public Transport in the City of Madrid

Gonzalo Fernández-Sánchez, Juan Ángel Terrón and Álvaro Fernández-Heredia

Abstract

This chapter is a vision of the path followed by EMT of Madrid, during 25 years, towards the sustainability, efficiency, and contribution to the air quality in the city, starting from a diesel fleet, until getting a fleet 100% of clean vehicles, mostly GNC, which is already a reality. It shows the evolution and the use, from the practical perspective of an operator, of all the technologies available at each moment (bio-diesel, bioethanol, hydrogen, electricity, natural gas, hybridization, dualisation, start-stop, catalysis, etc.) in Madrid, in a fleet of more than 2000 buses, more than 200 lines, and more than 400 million passengers per year, which makes this case an international benchmark. In addition, EMT is currently at the end of the transition to gas vehicles (CNG) and the implementation of urban electric mobility from the double perspective of the mobile material and the associated infrastructure needed, an essential case study towards sustainable public transport.

Keywords: sustainable transport, bus public transport, air quality, bus fleet, new bus technologies

1. Introduction

Worldwide, cities, with an extraordinary growth of their population, face a challenge in relation to air quality where traffic is a key element. Air pollution is the origin of a large number of deaths per year, greater than, for example, traffic accidents [1]. There are three paths towards a more sustainable transport: the rationalisation of the systems, the substitution of fossil fuels through renewable energies, and energy saving [2].

The energy dedicated to transport exceeds 33% of all used energy and 40% of the emissions. In the Community of Madrid, this amount is even greater due to the high population density and the importance of the service sector [2]. This makes transport one of the main causes to expose the urban population to air pollutants. Therefore, contributing to sustainable and efficient mobility in urban environments is essential for a better air quality and to reduce fuel consumption, loss of time, and waiting times, as well as polluting emissions.

In Europe, significant progress has been made in reducing the main pollutants in urban areas. However, a total of 23 countries exceeded the authorised daily maximum of PM10 particles in 2010 in at least one or more measuring stations [3].

In large urban centres, 66% of NO_x contamination, the most severe for people's health, is produced by road traffic. Buses on regular urban lines generate only 7.5% of this percentage, whereas private cars are responsible for 64.8% of said pollution [4].

Public transport is therefore essential to achieve the objectives of air quality, traffic decongestion, accessibility, and health in cities in terms of mobility [5]. Thus, authors such as Silvia Cruz and Kazt-Gerro [6] analyse the focus on public transport companies as the essential entities for the development of sustainability in the production chain and how public policies, management of economic resources, and the oil market restrict the possible strategies towards an improvement of the environment. In this chapter, we will dive deep into and analyse the strategies and implementation towards the sustainability of the urban bus service in the city of Madrid through the real experiences of the "Empresa Municipal de Transportes de Madrid" (EMT de Madrid).

2. Urban public transport on surface: EMT Madrid

EMT Madrid is a public limited company owned by the Madrid City Council that deals with urban bus transport in the capital. It is also entrusted with the management of other tasks related to mobility in the city, such as the city's tow truck, a part of the public car parks, the cable car, and the public city bike system, as well as an international consulting area.

The company was founded in 1947 to meet the demand for transport in Madrid. Currently, it is one of the most important urban transport companies in Europe in terms of fleet size, number of transported passengers, or technological innovation. More than 420 million passengers are transported annually and 95 million kilometres are covered in its 213 regular lines, of which 26 provide a night bus service. It employs more than 9300 workers and manages the city's 88 tow trucks, its 6 impound sites, and 24 of the city's car parks with almost 11,000 parking spaces [7]. Likewise, it has 5 operation centres from which the fleet is supplied and

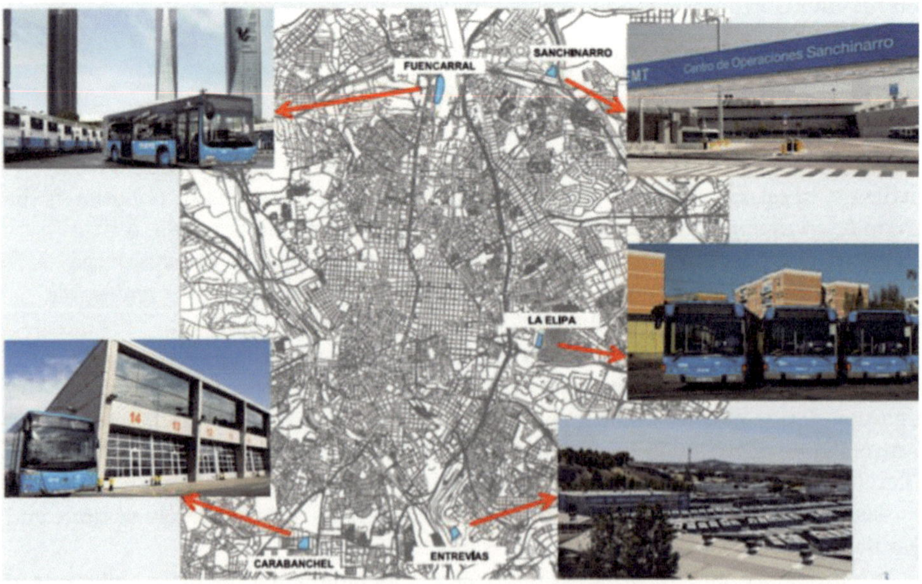

Figure 1.
Madrid and the distribution of its five operation centres.

maintained (**Figure 1**). More than half of this fleet is powered by compressed natural gas (CNG), and there are more than 80% of low emission buses. The average age of the bus fleet is 7.13 years (2018) and this figure is expected to continue decreasing in the coming years.

EMT has made a firm commitment in recent decades to be a sustainable company that takes care of the environment. The activity of the EMT makes it a key tool for sustainability and improvement of the Madrid environment.

3. Practical experience

According to UITP (International Association of Public Transport), approximately 80% of all public passengers worldwide are transported by bus [8]. Beyond the different developments in the bus system, there is no consensus on the most effective strategies [9]. In spite of everything, studies have been carried out relating to urban planning, mobility, air quality, and pollution by emissions and the connection of these elements with variables such as the average speed, the frequency of the bus service, and the bus system network to reduce these emissions [10]. Some of the studies that take into account fuel efficiency, traffic congestion, taxes on coal, and improvements in public transport conclude that a mix of strategic measures for public transport is the best option towards a sustainable environment [11]. Thus, the performance in infrastructure (either through BRT or exclusive bus lanes), electric vehicles or those running with new sources of energy such as natural gas and its positive impact on the drastic reduction of NO_x [12], the application of technology for service improvement [13] and user information or route optimization, international research, and development or service improvement are some of the most important strategies, which are developed in the bibliography [13].

It is clear that depending on the size of the city and the conditions of the environment, the strategy to be implemented will differ from one case to another. In the following, the real experience of EMT Madrid in the last decades is shown, acting in a transversal way simultaneously and synergic in all of EMT's activities. Thus, separate experiences are shown in operation and road infrastructure, mobile material (buses), the personnel and the facilities, and infrastructures, which are necessary for the operation (operation centres: refuelling, workshops, etc.) according to **Table 1**.

3.1 Operations and infrastructures in the street

3.1.1 Redesigning bus lines and sustainable mobility in the city

EMT within the scope of its competencies as an operator continuously performs a study of the bus lines in service. In 2018, more than 68,123,000 kilometres were studied, analysing the quality offered, the evolution of regularity, and the occupancy levels at peak hours; monitoring the basic service variables, the evolution of the urban fabric, and the demands of lines (individual and aggregate); and checking the programmed speed with the actual operating speed on a three- or four-month basis.

In this sense, the technology based on GIS (geographic information system) is being improved to improve the planning and exploitation of transport as well as the construction of the infrastructure for analytical models based on big data with new calculation of multimodal routes (bicycles–bus–on foot–parking lots). But it also strives to optimise the bus models of each line according to the environmental

Strategies for a sustainable transport-EMT case study	About operations and road infrastructure	• Redesign of bus lines and sustainable mobility in the city
		• Implementation of bus lanes and their control
	About the mobile material	• Establishment of vehicle acquisition criteria
		• Exhaust gas treatment
		• Start-stop system
		• Compressed natural gas (GNC) buses
		• Electric transmission
		• Fuel cell (hydrogen)
		• Hybrid buses
		• Electric minibuses
		• Recharging opportunity
		• Electric buses
	About EMT staff	• Training with driving simulators
		• Efficient driving
		• EMT staff transport
	About infrastructure design	• Energy saving and environmental certification
		• Adaptation of operation centres
		• New operation centre 100% electric
		• Public access charging points for electric vehicles

Table 1.
Strategies for sustainability in transport-EMT case study.

conditions of the area through which they pass or redesigning them ensuring the same service (speed, stops, and safety) producing the least possible contamination.

The ECCENTRIC project (CIVITAS H2020 funded by the European Union) aims to implement sustainable mobility solutions in peripheral areas of the city including public transport. EMT participates as a partner being responsible for the administrative coordination that directly impacts on the rethinking of the sustainable model. In this sense, EMT is also developing the MaaS (Mobility as a Service) project as an application that integrates and visualises the different mobility options to go from one point to another in the city at any time. International projects such as IMOVE are there to support the deployment and improvement of MaaS in Europe with innovative solutions.

The Dynamic Mobility Management project tries to implement this improvement for the planning and management of the urban bus service operations by assigning traffic and public transport that allows modelling scenarios and situations by analysing the impact different measures may have on mobility, on the transport network, and on the air quality of the city of Madrid, with the combination of own and external data such as telephone data or mobility apps. The functionality in the forecast of pollution episodes, the simulation of new scenarios, and the adaptation of supply to demand in a more agile way will allow greater flexibility and adaptability of the entire public transport system in Madrid.

3.1.2 Implementation of bus lanes and their control

BRT and exclusive bus lanes are a very common measure to prioritise and improve the speed of urban bus transportation. In this sense, EMT currently has 34,735 kilometres of road axes with bus lane separators, growing progressively and allowing the improvement in the quality of the service provided and the optimization of operation reducing fuel consumption and emissions that are linked to it. In 2018, in Madrid, a total length of more than 1.5 kilometres per direction was implemented with more than 395 separators and 260 beacons.

These lanes are at the same time controlled by EMT staff with contamination-free electric vehicles (SACE).

3.2 Bus fleet

The EMT fleet has undergone a continuous change in the last decades as shown by the example in **Table 2** that represents the evolution in the last 15 years, but that will be better understood when analysing each of the experiences and tests made in these years.

3.2.1 Establishing vehicle acquisition criteria

In the middle of the nineties, the division of "Definition of Bus Fleet" was created allowing the development of a list of conditions for the acquisition of buses with all the characteristics and conditions that must be fulfilled in order to minimise the energy consumption of vehicles and emission levels. In addition, the acquisition criteria are aimed at using clean energy and contemplating the quantification and assessment of emissions during the life cycle of the vehicle. This inclusion in the EMT documents was done long before the European legislation indicated it, not only taking into account the cost of acquisition. In this sense, EMT is integrated in the working groups related to bus fleet and the purchase of buses with the UITP with the idea of integrating and creating synergies sharing good practices at an international level.

3.2.2 Exhaust gas treatment

The bus fleet has been mainly diesel until 2016, so the emission reduction focused on finding alternative fuels and on changing the existing buses of the EMT

Año (31-12)	Gasoleo	Dual Fuel	GNC	Etanol	Transm. eléctrica	Electricos	Hidrógeno	Hibridos	Flota	Flota alternativa
2004	1789		155		20		4		1968	179 / 9,10%
2005	1802		165		20		3		1990	188 / 9,45%
2006	1793		201	5	20		3		2022	229 / 11,33%
2007	1647		351	5	20	10			2033	386 / 18,99%
2008	1634		381	5	20	20			2060	426 / 20,68%
2009	1670		393	5	12	20			2100	430 / 20,48%
2010	1603		465	5	3	20		4	2100	497 / 23,67%
2011	1415		651	5		20		4	2095	680 / 32,46%
2012	1234	1	723			20		22	2000	766 / 38,30%
2013	1088	1	767			20		27	1903	815 / 42,83%
2014	1090	3	767			20		27	1907	817 / 42,84%
2015	1091	3	767			20		27	1908	817 / 42,82%
2016	1068	3	799			18		27	1915	847 / 44,23%
2017	928	3	1019			23		52	2025	1097 / 54,17%
2018	629	3	1334			36		48	2050	1421 / 69,32%

Table 2.
EMT fleet evolution 2004–2018.

fleet. Thus, in order to reduce the contribution of urban transport to NO_x pollution, in addition to using fuels that produce it in smaller amounts, it became necessary to implement exhaust gas treatment systems in buses that were already in service, using oxidation and reduction catalysts to eliminate both the particles and the NO_x.

Between 2012 and 2014, 510 catalysts were installed in diesel buses to improve their emissions. These elements make buses that were originally approved according to the Euro III emission regulations to emit exhaust gas at levels lower than those seen in the Euro V Standard, with which the Low Emissions Zone of Madrid was fully served by low-polluting buses.

3.2.3 Start-stop system

The start-stop systems achieve a saving in fuel consumption and a reduction of polluting emissions, stopping the combustion engine when it is idling due to a stop of the vehicle. The system is activated at traffic lights and at passenger stops, as well as in other traffic conditions that involve stopping the vehicle for a certain time. They allow, therefore, having the heat engine turned on only when the bus is in motion.

A bus, running in Madrid, travels an average of 200 km per day and performs about 50 stops (more than 5 seconds) per hour, standing in each of them about 28 seconds during 15 hours a day, which accounts for almost 40% of the total work time. Since the fuel consumption at idle is 0.5 g/s, a fuel saving of 10.5 kg of diesel per day is achieved, that is, an 8% decrease, similar to the reduction in pollutant emissions [14, 15].

3.2.4 Compressed natural gas (CNG) buses

EMT has been one of the first companies in Europe to use CNG as a fuel for urban buses, receiving the National Environmental Award for this strategy in 1998, although other companies such as New York, Melbourne, or Paris are also starting on this way.

In 1994, EMT began its experience in the use of CNG as fuel by participating in the ECOBUS Project of the European Union, under the umbrella of its THERMIE programme. After the success of this project and the results obtained by being one of the cleanest fuels that exist, being compatible with diesel technology, being assured of its regular supply, having a stable price, and being an alternative to oil, EMT of Madrid decided, as of 1995, to introduce the use of CNG as a new fuel into its fleet. In that year, the first compressed natural gas station with a capacity for 20 buses was implemented, charging them in about 3 hours. Since then, the number of buses in the fleet powered by CNG has gradually increased and, consequently, the size of the charging facilities, which have grown in power, efficiency, and speed of charging. Today, the refuelling is carried out by charging stations using powerful compressors that supply fast filling stations in 3 minutes (**Figure 2**).

As it is currently the best alternative to diesel from the environmental and economic point of view, EMT, from 2010, took the decision to adopt CNG as the basic fuel of its fleet, so that all buses that were acquired should use alternative energy to diesel and be CNG, hybrid, or electric.

At the end of 2018, more than 80% of the fleet was considered green (this is those that comply with the Euro V standard or higher, CNG, electric, or hybrid vehicles). It is planned that by 2020 the entire EMT fleet will be CNG, as well as a small number of electric and hybrid vehicles. In all the operation centres, CNG refuelling stations are already available, with a centre entirely designed for CNG buses (Sanchinarro) (**Figure 3**).

Figure 2.
First GNC bus in 1994 and first charging station of 20 GNC buses.

Figure 3.
New bus fleet.

EMT's use of this type of buses has meant a reduction of 75% in NO_x emissions and 95% in particles, while the advance of diesel technology has allowed an average reduction of 55% NO_x emissions and 80% of particle emissions in the same period.

3.2.5 Dual-fuel project

Compressed natural gas (CNG) has economic and environmental advantages over diesel. For this reason, EMT has been interested in a dual-fuel technology, which allows natural gas to be used simultaneously with diesel fuel, in diesel buses [14, 15].

The dual-fuel (or shared combustion) technology can be defined as the simultaneous combustion of two fuels; in this case, natural gas is used in combination with diesel to operate the diesel engine. In urban transport, percentages of 50% diesel and 50% natural gas can be reached. Once the modification has been made, the engine can operate in dual-fuel mode or exclusively with diesel, but in no case with natural gas exclusively.

The use of the mixture of natural gas and diesel significantly reduces the emissions of polluting gases, which depend on the engine's operating conditions and the substitution levels of diesel that are reached. Assuming typical substitution levels of 50%, CO_2 reductions are between 10% and 15%, particle reductions can reach 50%, and NO_x levels are between 35 and 65% [16].

In addition, by replacing a part of diesel with CNG, which is cheaper, there is an economic benefit that allows the return of the investment and makes the project economically viable.

3.2.6 Electrical transmission

In the year 2000, a fleet of 20 electric transmission vehicles was put into service to run in the city centre. These were vehicles of medium length (9 m) that had an electrical transmission consisting of a generator, power electronics, and an electric motor, in addition to the auxiliary elements that this entails, instead of the traditional mechanical kinematic chain (clutch and gearbox). However, the energy is still supplied by a diesel thermal engine, does not carry energy, and does not recover braking energy.

The performance of the electric kinematic chain was superior to that of the mechanical chain, which results in savings of fuel and emissions of 10% compared to a conventional vehicle of the same size that performs the same type of service. Likewise, the comfort of the passengers improves when increasing the smooth running of the electric motor (**Figure 4**).

3.2.7 Fuel cell (hydrogen)

EMT participated during the 2003–2005 period in the CUTE/ECTOS and CITYCELL projects, consisting of the start-up of electric buses equipped with a fuel cell. This type of buses uses hydrogen as the primary energy vector, generating the electric energy necessary for traction and producing water vapour as the only by-product [14, 15].

When participating in the two projects, EMT Madrid was the only company that had, simultaneously, 4 fuel cell buses in service.

Project Cute: It consisted of 3 Mercedes Benz buses with a 205 kW fuel cell working in the full power mode, that is, producing electricity at the time it was needed, without accumulation and working on a transitory basis (**Figure 5**).

Project Citycell: It consisted of 1 IVECO bus, with a 62 kW fuel cell, functioning as a hybrid vehicle, with storage batteries and working in stationary mode (**Figure 6**).

It was shown that the fuel cell technology is viable in urban transport, with zero emissions of exhaust gases and very low noise pollution, but also showing that it is necessary to increase the autonomy and reliability of vehicles and charging stations, as well as how to create a hydrogen distribution network.

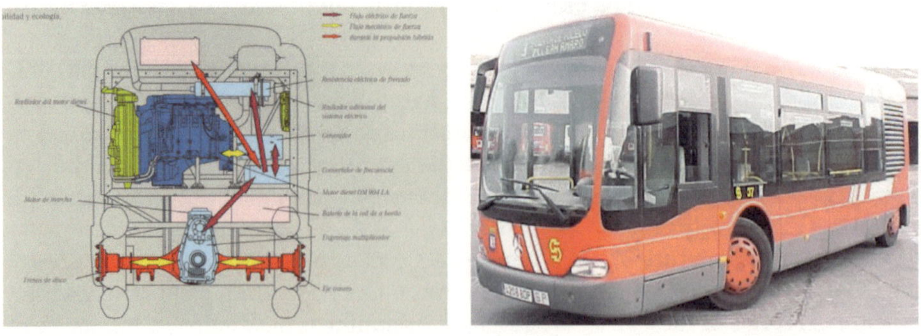

Figure 4.
Electrical transmission bus.

DOI: http://dx.doi.org/10.5772/intechopen.90102

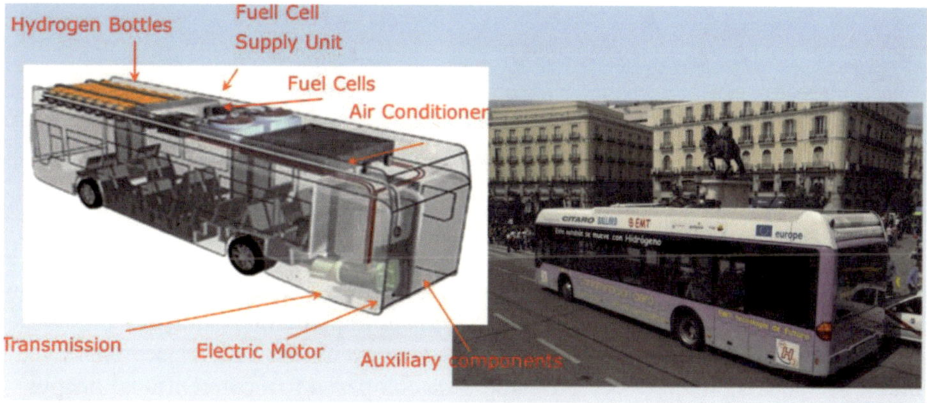

Figure 5.
Hydrogen bus.

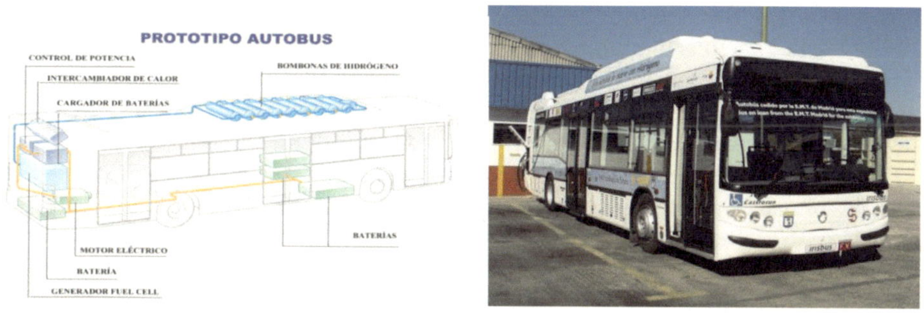

Figure 6.
Hydrogen bus, Citycell.

3.2.8 Hybrid buses

The need for sustainable transport in today's society is beyond doubt, as well as that electricity will be the driving power of the future. But, with the current state of energy accumulation technique, buses have problems of autonomy to perform the usual daily service if pure electric propulsion is used, having to resort to intermediate recharges in their working day.

Due to this, the use of buses with hybrid propulsion, by means of the use of thermal and electric motors, and with energy accumulation systems, has become general and will be essential during the transition period from buses with current thermal engines to the purely electric buses of the future.

With hybrid propulsion, it is possible to run with electric traction, energy sourcing from the accumulated energy for a certain time, recharging these accumulators when necessary through the thermal engine, thus achieving the required autonomy, fuel savings, and a reduction of polluting emissions.

In urban transport, depending on the employed control strategy, hybridisation can be used as a tool to improve air quality in the city, allowing the operator to decide when the bus is running in pure electric mode (for example, in the ZBE), thus contributing to the local reduction of emissions (**Figure** 7).

3.2.9 Diesel hybrid

In 2016, EMT acquired 30 hybrid diesel buses, 17 from MAN, equipped with ultra-capacitors, and 13 from IVECO, both with non-pluggable batteries. These

Figure 7.
Hybrid buses.

vehicles are used to run in specific lines in the east of the capital and are included in the European project ECCENTRIC, which aims to implement sustainable mobility systems integrated with urban planning. The savings in fuel and, consequently, emissions and CO_2 are between 30% and 35% compared to a conventional propulsion vehicle.

3.2.10 GNC hybrid

EMT asked the bus manufacturers Castrosua and Tata Hispano to develop a hybrid bus powered by natural gas and thus emerged two models of CNG hybrid buses of which EMT has acquired 23 units, 10 from Tata Hispano and 13 from Castrosua, being the first CNG hybrid buses to be marketed in Europe.

The consumption and the CO_2 emissions have been reduced between 28% and 35%, in comparison with the values of a CNG bus and up to 70% emissions of NO_x and PM with respect to a diesel vehicle.

3.2.11 Converted buses

EMT Madrid participated, together with TMB Barcelona and EMT Valencia, in the Electrobus Project, with financing from the Institute for Diversification and Energy Saving (IDAE), consisting of the conversion of diesel buses into hybrid diesel-electric buses. In 2011, a conversion of 4 Iveco CityClass Euro III diesel buses was carried out on electric diesel hybrid buses, which have reduced their diesel consumption and their emissions by 18% compared to conventional propulsion vehicles.

The transformation consisted in the elimination of the gearbox, the installation of a 180 kW electric generator driven by the original thermal engine, the incorporation of two tandem electric traction motors, each with 67.5 kW of power, and the incorporation of control and auxiliary elements such as inverters, radiators, or a steering pump.

As energy storage elements are installed, 6 units of supercapacitors, with a total voltage of 750 volts and a capacity of 10 farads, are put on the roof area.

3.2.12 Biofuels

In addition to all the above, EMT has carried out different projects and pilot tests with biofuels (reduced percentages of biofuels with diesel) from the Bio-Bus prototype (1997–1999), and the following projects and trials were carried out:

The BD5 biodiesel trial was developed from November 2003 to March 2005 with four buses equipped with a Euro III engine running over 260,000 kilometres. The biodiesel used was a BD5 with 5% methyl ester derived from sunflower oil. There were no problems in the bus injection system, having correct values of the analysis of filters and oils used. There is no significant increase in consumption of buses that use diesel as fuel (**Figure 8**).

DOI: http://dx.doi.org/10.5772/intechopen.90102

Figure 8.
Biodiesel bus: Bio-bus.

Project Biodiesel EHN 100%: the specifications of Euronorm 14,214 and RD 1700/2003 were followed. It was used pure without mixing with diesel oil, from vegetable oils of first use and a mixture of different types mainly soy, sunflower, rapeseed, and palm. The tests were made between June 2005 and October 2006 on 6 vehicles: 4 Euro II (two Mercedes O/405 and two Iveco CityClass Cursor all from 1998) and 2 Euro III (Iveco CityClass Cursor 2002). They made more than 300,000 kilometres, not seeing an increase in breakdowns in the injection and power systems compared to the rest of the buses of the same model and year. There was a slight increase in fuel leakage failures in the Mercedes model (7 years old).

Project Biodiesel CLM: 20% supplied by Biodiesel Castilla La Mancha obtained by mixing 20% of methyl esters of vegetable oils from the recycling of used oils with 80% conventional diesel. The test was carried out from August 2005 to September 2006 in 6 buses: 3 MAN NL-263F of 2003 and 3 IVECO CityClass Cursor of 2004. These buses made more than 300,000 kilometres, and there were no breakdowns in the systems, injection, or supply.

Biodiesel test invariable proportion: a third test was carried out on a MAN Nl-263F of 2003 from sunflower oils of first use, testing the benefits and the consumption with biodiesel varying the proportion between 20 and 100%. This bus covered more than 40,000 km between January and December 2005 without specific problems, or breakdowns in the power system, although there was an increase in consumption and power decrease that increased with the proportion of non-linearly used biofuel (**Table 3**).

Biofuel in the entire fleet: in October 2006, the use of large-scale biofuels was started in all diesel cars in a depot of EMT (Fuencarral A) with 20% biodiesel and mixing directly in the EMT storage tanks. The fleet's study objects were 150 IVECO CityClass Euro II, 30 IVECO CityClass Cursor Euro III, and 30 Mercedes O/405 N (Euro II).

3.2.13 Electric minibuses

Since 2007, EMT has had 20 Tecnobus electric minibuses, model Gulliver, that provide service in two lines that run through the narrow streets of the historic centre of Madrid, within the ZBE. During its journey, the environmental conditions are not altered by emitting any polluting gas and it only produces the dull rolling noise of the tyres on the asphalt. These small vehicles, 5.5 m long, work with the energy stored in two Ni/NaCl accumulator modules, 72 kWh at 85 V. The weight of each module is almost 300 kg and they are located in the rear of the vehicle.

Types of biodiesel		Increase of average consumption
BD5		1.54%
CLM 20%		4.50%
EHN 100%		7.50%
Variable	100% bio	9.90%
	50% bio 50% diesel	6.50%
	30% bio 70% diesel	6.30%
	20% bio 80% diesel	4.40%
B20 (entire fleet)		3.54%

Table 3.
Types of tested biodiesel and increase in average consumption (adapted from Terrón [17]).

The electric motor that moves the bus, front-wheel drive, has a maximum power of 27.2 kW, reaching a maximum speed of 32 km/h. This achieves a commercial speed of 6 km/h, a satisfactory figure given the characteristics of the streets through which it runs. This has saved about 6000 litres of diesel fuel per minibus per year, which would have been consumed in the case of this transport with conventional vehicles, with the consequent reduction pollutants and greenhouse gases.

There is also another electric minibus, Breda model ZEUS, for internal and institutional transport.

The electric minibuses have been replaced in full between 2018 and 2019 by 18 electric minibuses of the Wolta model of the manufacturer Microbuses de Lujo SL (**Figure 9**).

3.2.14 Recharging opportunity systems

Since the electric buses, which currently exist in the market, do not have the required autonomy, it is necessary to replace them in the middle of their working day with others, that are already charged, or to use intermediate charging systems during their service called recharging opportunity systems. These recharge the vehicle partially, on the line itself, usually in its terminals. In this way, by means of

Figure 9.
Electric minibus.

successive recharging during their working day, their autonomy increases indefinitely.

The recharging opportunity can be carried out by conductive charging, by contact, generally with a pantograph that is installed in a more or less visible structure; or with the newest and least visible system of recharging-by electromagnetic induction, without physical contact between the vehicle and the charging terminal, carried out by means of a primary coil on the underside of the bus and a secondary coil under the pavement of the road. During the time the vehicle remains at the final destination of each trip, the batteries are charged without any other human intervention than placing the vehicle on the loading area. This allows the use of a safe and fast charging system, eliminating visual pollution in the city and extending the autonomy of the bus indefinitely. It is a very interesting system for specific lines that enter certain sensitive areas of the city, since it enables the electric mobility of the bus without any autonomy problems.

EMT has recently (in October 2017) implemented a line of electric buses with recharging opportunities by induction. Previous experiences allow ensuring a system efficiency of 95%, leading to savings higher than 15%, since there will be no empty trips due to the lack of autonomy of the vehicles. In addition, due to the recharging opportunities, the number of batteries installed on board of each bus can be lower, having a lower consumption per unit. Thus, there is already one line (line 76 of Madrid) that consists of high-power recharging opportunities on a route that connects an outer neighbourhood of Madrid with the city centre, where the existing buses are replaced by zero-emission vehicles whose retrofit stems from an investment in R&D converting a hybrid bus with CNG to 100% electric vehicle with induction charging, little on-board energy, and little autonomy, which is charged in each terminal (**Figure 10**).

3.2.15 Electric buses

EMT believes that the future of urban transport, in the long term, passes through electric mobility, which will be the only one capable of making cities environmentally sustainable. In order to achieve the electrification and decarbonisation of urban transport in one decade, it is necessary to start implementing that mobility from today on.

The advantages of using the electric vehicle can be easily grouped into three areas: energy efficiency, through a more rational use of the consumed energy; environment, with a substantial improvement in the global pollution emitted "from the well to the wheel" and a total elimination of local pollution; and demand management, through greater efficiency of the electric system, reducing the energy dependence on fossil fuels and using the off-peak hours for charging. However, the current state of the art regarding the autonomy of the buses, the charging of the batteries, and the state of the electrical distribution networks, as well as the cost of vehicles and facilities, are barriers that make the road to the future neither easy nor fast.

EMT is starting on this road to the electro mobility of its fleet with the aforementioned recharging opportunity project, beginning with the implementation of 12 metre long electric buses, with a strategy of growth of the electric fleet in Madrid that includes actions in a term of about 6 years, so that by 2020 there are going to be about 80 electric buses, a figure that will increase to 250 by the end of that period.

As for standard electric buses of 12 metre length, EMT has carried out tests with the main bus models of these characteristics that are being marketed in Europe. The brands are IRIZAR, BYD, FOTON, EURABUS, EVOPRO, SOLARIS, and IVECO, all of them with a low floor. With the exception of IRIZAR, whose batteries are made

Figure 10.
Bus charging by induction and necessary installations.

of molten salts, the rest are equipped with lithium-ion batteries. Although each one has a different storage capacity, they all exceed 200 kWh of stored energy; only the IRIZAR model reaches 376 kWh.

From the tests carried out, it can be deduced that current electric buses have an energy consumption of between 1.5 and 1.7 kWh/km, so with the current state of the art, they cannot yet compete with the autonomy of other types of energy, but they are increasingly approaching that at a high speed.

EMT has acquired 15 electric IRIZAR buses that will serve as initial pilot fleet to gradually increase the number of "zero emission" buses running in Madrid. Thus, in February 2018, the first electric charging station was commissioned in the Fuencarral Operation Centre with 80 kW chargers and its own transformation centre. It is foreseen that in 2019 up to 35 standard electric buses will be integrated, currently finishing the necessary electrical installation by means of an electrical split achieving the necessary power, thereby taking advantage to test all the available technologies (charging by pantograph, chargers, and intelligent charging) and anticipating the execution of the new project that is already designed and planned to be built in 2020 from a 100% electric operation centre with a capacity for 330 buses (**Figure 11**).

3.3 Human resources

3.3.1 Training with driving simulators

EMT owns four driving simulators that reproduce EMT's existing bus models as well as simulate the operation and traffic (private, bicycle, etc.) that can occur in the city. These hours of practice in simulators contribute significantly to energy

Figure 11.
Electric standard bus.

savings and reduce the emission of gases and noise in the process of training driving staff. Training for all EMT employees in the year 2018 amounted to a total of 200,697 hours, taking into account that most were focused on driving staff.

3.3.2 Efficient driving

As it not only affects the operation and exploitation of the urban bus service or the rolling stock and infrastructures, but also the human component, EMT is currently working on improving the driving efficiency of its more than 2000 buses with the dual objective of reduction of polluting emissions and fuel consumption, as well as a certain improvement in the maintenance and conservation of the vehicle.

The EFIBUS project (first project in Madrid with the Innovative Public Procurement method) therefore seeks to reduce emissions, improve safety and travel experience, and offer detailed information to know the vehicle's condition. Thus, the technological solution focuses on an on-board module that stores and processes the relevant information (acceleration/abrupt braking and speed) informing the driver, in real time, of its degree of efficiency in driving and sending the control centre data for further analysis. The objective is to try to compare the same conditions to the driver with himself in the same conditions and a continuous improvement plan focused on each individual.

In addition, the module allows controlling the switching on and off of the bus's own systems such as lighting, air conditioning, ticketing systems, etc., which also allows a reduction in the energy consumption on the whole. The incorporation of these systems, together with a policy of training and continuous assessment of drivers in efficient driving, allows to achieve estimated reductions of up to 11% in the short term (3% in a sustained manner over time).

3.3.3 Transport of EMT staff

Since 2018 EMT also has a Transportation to Work Plan as a commitment between the management of the centres and the company's staff, which is characterised by the rationality in the journeys generated by the work activity. Thus, there are staff buses and auxiliary electric, hybrid, and CNG vehicles. But additionally, the recurrent trips that are made every day because of work are sought to redirect them towards socially and environmentally sustainable modes of transport. The Air Quality and Climate Change Plan of the City of Madrid includes the Sustainable Labour Mobility Plans and specifies actions to be carried out: "development of sustainable mobility plans in companies and public bodies, as well as in business areas of the capital". Currently, this Transportation to Work Plan is implemented at the headquarters, beginning its implementation in the operation centres and other EMT units. For this, a survey asking the staff has been carried out with a total of 646 responses (about 9530 workers at that time). The modal distribution, the geographical distribution of the origin of the trips, and the time used to arrive in the case of headquarters offices (with 1357 workers) are shown below. As a support, an advisory group and a car sharing platform have been created, as well as a new distribution of parking spaces giving priority to ecological vehicles, providing transport and travel cards for the use of other ways of sustainable public transport (for example, Madrid's public bicycle), as well as considering the family situation.

On the other hand, for the internal service of the company, there is a fleet of auxiliary vehicles for tourism within which, since 2015, there are 5 Renault ZOE electric cars equipped with 22 kWh lithium-ion batteries and a 65 kW electric motor, with autonomy close to 150 km and a maximum speed of 135 km/h (**Figure 12**).

Figure 12.
EMT auxiliary vehicles to support the service.

3.4 Installations and infrastructures

In fixed installations and buildings, from the design stage of the construction project, whenever possible, it must be taken into account that they are energy efficient, use renewable energy, and use efficient recycling.

With this concept arose the operation centres of Carabanchel, for diesel vehicles, in 2007, and Sanchinarro, built for vehicles of compressed natural gas, following the decision of EMT in 2010 that the new buses had to be predominantly of this fuel. The Sanchinarro Center was designed and built to employ construction techniques that are not aggressive for the environment, generating little waste and maximising recycling (**Figure 13**).

3.4.1 Energy saving and environmental certification

The energy saving of the centres has a direct impact on the consumption of gas and electricity, as well as on the level of polluting emissions. For this reason, an exhaustive control of the correct operation of the facilities is carried out to try to achieve continuous energy efficiency.

The roofing floors of Carabanchel and Sanchinarro are equipped with solar, thermal, and photovoltaic panels, to reduce the consumption of fossil energy and, consequently, the pollutants emitted into the atmosphere, thus contributing to improve the air quality of the city.

Respect for the environment has been of great interest in the design of buildings, as evidenced by the fact that natural gas is used as fuel for all the Centre's machinery, to filter the ventilation air, which is returned to the exterior after recovering part of the used energy to collect the washing water from the buses for their subsequent recycling and reuse, and to increase the green areas, having a landscaped part of the roof and vertical gardens.

EMT has the management systems ISO 9001, 14,001, and EMAS that contribute to the sustainable development of the company as a reference in the process of awareness, respect, and conservation of the environment that surrounds us, having as objectives the minimization of environmental impacts and the generation of an image committed to preserving the environment as an example to follow. Thus, controls are carried out on the impacts on the soil, water, and atmosphere, acting preventively and correctively according to the implemented management systems.

Figure 13.
View on the Sanchinarro operation Centre built in 2010.

In this sense, clauses of social content related to responsible management in the supply chain are also included in the specifications and contracts.

3.4.2 Adaptation of operation centres

All the actions on the exploitation and training and, mainly, on the bus fleet would not be possible without the suitable adaptation of the different operation centres.

To achieve greater efficiency in the management of the centres, the operation and maintenance of a more sustainable fleet as well as the improvement in the working conditions of the employees of the EMT have carried out a series of actions as those listed that highlights the installation and improvement of CNG stations (except for Sanchinarro, which was already 100% CNG, all the centres have been adapted for the operation of a 100% CNG fleet) during the years 2017, 2018, and 2019, as of electric recharging points for buses (in Carabanchel and Fuencarral) in those same years. In addition, actions have been taken in training (improvement of simulators) in 2018 and in the recirculation of water in the washing stations as well as in the pre-treatment of water.

3.4.3 New 100% electric operation centre

The recently completed project (ended 2018) of the new operation centre La Elipa, which will be executed according to the quality, finishing, design, and installation standards of Carabanchel and Sanchinarro, will begin its execution by 2020 with an execution deadline of 18 months. The operation centre is conceived with sustainable criteria, with plant facades for its landscape integration within the environment located in a consolidated residential neighbourhood. This centre shows that in the future 100% of the fleet will be constituted by electric buses. This project has been carried out with BIM technology to facilitate work monitoring, management, and subsequent operation of the centre. It has 313 places for buses of 12 metres and 20 places for articulated buses, 32,200 square metres bus parking

Figure 14.
New 100% electric operation Centre La Elipa (project finished).

platform, and a workshop building with a floor area of 6200 square metres. The underground parking of a basement can accommodate 347 spaces.

The building has been designed with almost zero energy balance criteria (ZEB building), through the use of recycled materials for construction, increased energy efficiency, the obtaining of renewable sources, and the provision of elements that favour the reduction of emissions of greenhouse effect. It has a photocatalytic cover (34,000 m^2) with a decontaminating effect that will reduce the pollution of the environment (elimination of NO_x polluting oxides) and that will generate an important energy saving, and a vertical garden of 1100 m^2 of vegetated air purifying modules, with vegetal facade translucent of 800 ml of climbing species and 400 m^2 of landscaped roof. Additionally, it has a photovoltaic installation with more than 7200 panels installed on the roof that will generate up to 2 MW of energy (**Figure 14**).

3.4.4 Public access points for charging electric vehicles

Likewise, compatible with this strategy of global electric mobility in Madrid, EMT is innovating the existing recharging systems in the facilities it manages in the city (car parks, operations centres, etc.), installing fast charging points for public use, such as the case with those implemented during 2017 and 2018 in several car parks, with the participation of the Madrid City Council, which has a dock for simultaneous rapid loading of four vehicles. At this moment, there are already 75 slow charge points and another 12 fast charge points. In the car parks of residents managed by EMT, it is sought to provide the opportunity to have electric vehicles supporting electric charge in them. Additionally, an electric station for public use is being proposed for the new operations centre of La Elipa.

4. Conclusions

All the aforementioned actions have contributed to the drastic reduction of NO_x and PM emissions of the EMT fleet in the last decade. The challenge of EMT and the city is to continue with the achieved downward trend. But while the electro mobility plans are being carried out, EMT has already completed its transition strategy from diesel to CNG as a real alternative to traditional fossil fuels, for economic and environmental reasons. Thus, by 2020, all urban buses in the capital will be powered by CNG, with the exception of those previously mentioned, leaving diesel as a fuel of the past (**Figure 15**).

By taking all these measures, our users can enjoy EMT's mobility offer and move around Madrid using the least amount of energy necessary for transport, using the least polluting techniques and technologies and having a company that is a European reference in the urban transport sector.

According to the results of EMT, in the following figures, the NO_x and CO_2 emissions per kilometres of some of the mentioned technologies can be seen, their cost per kilometre with a focus on the cost of the fuel in each case. There is still a way to go, but what has been achieved so far is, of course, a successful transition towards CNG and an initiated transition towards the future of electro mobility (**Figure 16**).

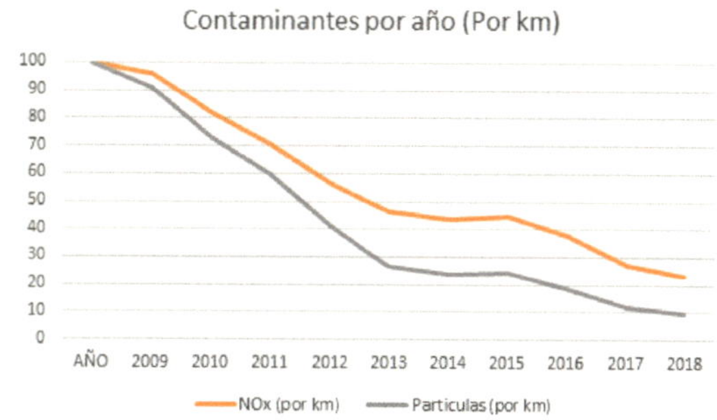

Figure 15.
Evolution of NO_x and PM emissions according to the EMT fleet.

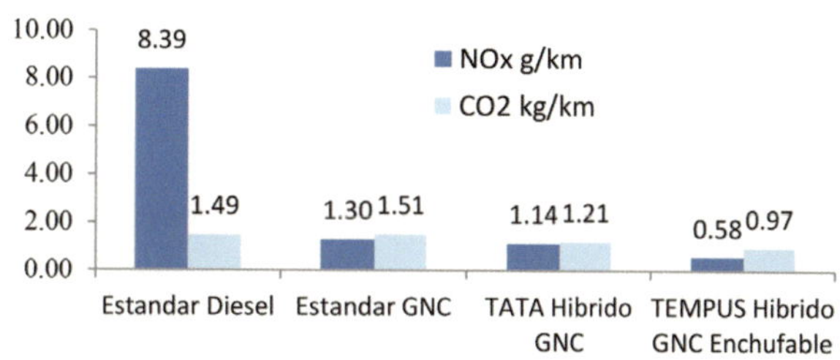

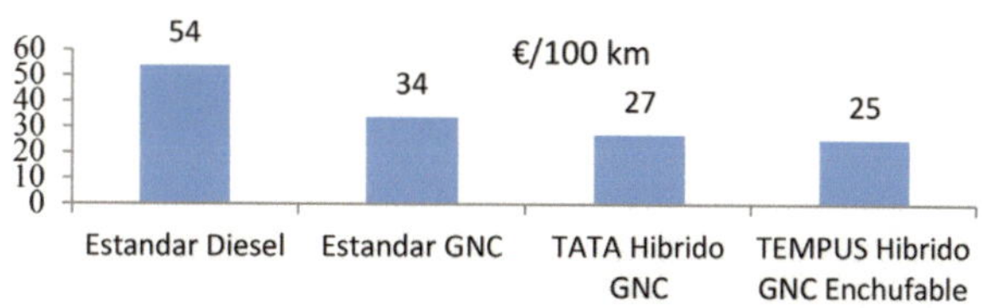

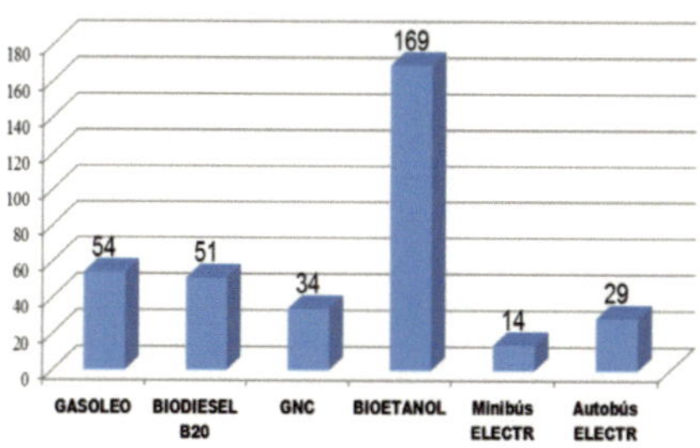

€/100 km	GNC	Gasóleo	%
Coste combustible y compresión	34	54	-37 %
Coste de mantenimiento	53	51	4 %
Coste adquisición	38	33	15 %
Total	125	138	-9 %

Figure 16.
Own results of different technologies and impact on costs and emissions.

Author details

Gonzalo Fernández-Sánchez*, Juan Ángel Terrón and Álvaro Fernández-Heredia
EMT de Madrid, Spain

*Address all correspondence to: gonzalo.fernandez@emtmadrid.es

References

[1] Palomo J. Moving people Chaning minds (91-108). In: Guía de Eficiencia Energética en la Movilidad y el Transporte Urbano. Móstoles, Madrid: Fundación de la Energía de la Comunidad de Madrid y Comunidad de Madrid; 2014

[2] de Madrid C. Guía de Eficiencia Energética en la Movilidad y el Transporte Urbano. Móstoles, Madrid: Fundación de la Energía de la Comunidad de Madrid y Comunidad de Madrid; 2014

[3] Aparicio F. Movilidad urbana y electromovilidad (13-31). In: Guía de Eficiencia Energética en la Movilidad y el Transporte Urbano. Móstoles, Madrid: Fundación de la Energía de la Comunidad de Madrid y Comunidad de Madrid; 2014

[4] Available from: https://www.sciencedirect.com/science/article/abs/pii/S0739885908000048

[5] Currie G, Rose J. Growing patronage–challenges and what has been found to work. Research in Transportation Economics. 2008;**22**:5-11

[6] Silvia Cruz I, Katz-Gerro T. Urban public transport companies and strategies to promote sustainable consumption practices. Journal of Cleaner Production. 2016;**123**:28-33

[7] EMT de Madrid. Informe de Gestión 2018. Madrid: EMT de Madrid; 2018

[8] Cascajo R, Monzon A. Assessment of innovative measures implemented in European bus systems using key performance indicators. Public Transport. 2014;**6**:257-282

[9] Currie G, Wallis I. Effective ways to grow urban bus markets–A synthesis of evidence. Journal of Transport Geography. 2008;**16**:419-429

[10] Ambarwati L, Verhaeghe R, van Arem B, Pel AJ. The influence of integrated space-transport development strategies on air pollution in urban areas. Transportation Research Part D: Transport and Environment. 2016;**44**: 134-146

[11] Macario R, Jara-Diaz S. Growing patronage: Challenges and what has been found to work. Research in Transportation Economics. 2008;**22**:12-15

[12] Hidalgo D, Muñoz JC. A review of technological improvements in bus rapid transit (BRT) and buses with high level of service (BHLS). Public Transport. 2014;**6**:185-213

[13] Fernandez-Sanchez G, Fernandez-Heredia A. Strategic thinking for sustainability: A review of 10 strategies for sustainable mobility by bus for cities. Sustainability. 2018;**10**(11):4282

[14] Hensher DA. Future bus transport contracts under a mobility as a service (MaaS) regime in the digital age: Are they likely to change? Transportation Research Part A. 2017;**98**:86-96

[15] Terrón JA. Case 1: The experience of the Madrid Municipal Transport Company (156-171). In: Air Quality in Cities. A Global Challenge. Madrid: Naturgy Foundation. Coord. Xavier Querol; 2018

[16] Brachetti J. 2010. Dual–Fuel the Best Fuel in the Most Efficient Engine, NGVA Europe Position Paper, Mayo 2010. Available from: https://www.hibridosyelectricos.com/articulo/libros/electromovilidad-transporte-urbano-colectivo/20140714182255007515.html

[17] Terrón JA. Autobuses eléctricos e híbridos en EMT de Madrid (35-47). In: ASEPA, 2014. Madrid: La

electromovilidad en el transporte urbano colectivo. Coordinador: Antonio Mozas Martínez. Monografías ASEPA. Asociación Española de Profesionales de Automoción; 2014

Section 2

Biofuels

Chapter 3

Feedstocks for Advanced Biofuels

Enrique Espí, Íñigo Ribas, Carlos Díaz and Óscar Sastrón

Abstract

Advanced (also known as second-generation and third-generation) biofuels must comply with sustainability criteria related to the feedstock used (not competing directly or indirectly with food or feed crops) and to their greenhouse gases emission reduction. These fuels must reach at least 0.2% of transport energy used in Europe in 2022, 1% in 2025, and 3.5% by 2030. In this chapter, sustainable feedstocks that could be used for producing advanced biofuels are reviewed: energy crops grown on marginal land; wastes and residues—agricultural, forestry, food, municipal solid waste, and other organic wastes and residues; and novel feedstocks—such as aquatic biomass (macroalgae and microalgae).

Keywords: advanced biofuels, energy crops, municipal waste, agricultural waste, forestry waste, biomass, algae

1. Rationale and regulation in Europe

The European Commission has recently published a document defining the European strategy to fight against climate change—A Clean Planet for all [1]. Recognizing that climate change represents threat to the planet, the Commission has set the goal, in accordance with the 2015 Paris Agreement, of keeping global warming below 2° above pre-industrial levels and pursuing efforts to limit it to 1.5° by 2050. Bioenergy is a key factor to achieve this goal, but reducing greenhouse gas emissions is not its only advantage: bioenergy can also contribute to energy security, create thousands of new jobs in Europe, especially in rural areas, and impulse the growing European bioeconomy, with synergies with other sectors, such as food, feed, bio-based materials, and chemicals.

Sustainable biofuels for transport, subject to the updated sustainability criteria currently proposed by the European Commission [2], are one important part of the bioenergy sector as they are easily deployable using existing transport infrastructure being the only near-term alternative to fossil fuels for some applications, such as marine or aviation fueling.

The Renewable Energy Directive (RED) [3], revised in 2018 (RED II) [2], the Fuel Quality Directive (FQD) [4], the "ILUC Directive" [5], and "the ILUC Commission Delegated Regulation" [6] define biofuel's sustainability criteria in the EU to ensure that they are produced in a sustainable and environmentally friendly manner. Current legislation requires a 7% cap on the contribution of conventional biofuels, including biofuels produced from energy crops to count toward the targets in 2020 and 2030. The RED II directive also sets as a binding minimum of 0.2% target for advanced biofuels by 2022 and 3.5% by 2030 (**Figure 1**). Finally, the directives harmonized the list of feedstocks (Annex IX) for the production of advanced biofuels across the EU. Those can be considered to count double (i.e., to be twice

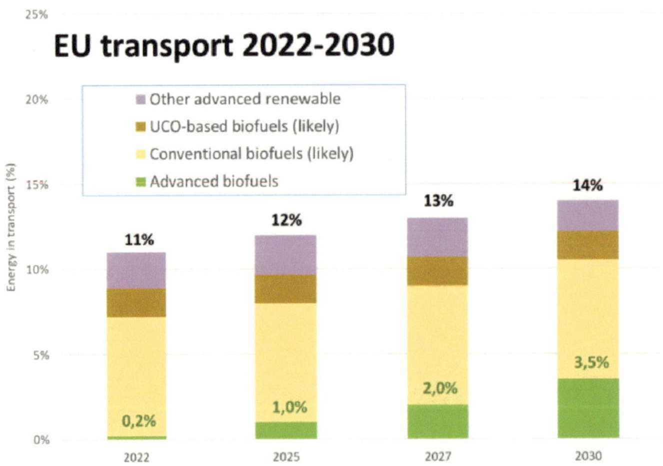

Figure 1.
RED II mandates about renewable energy in transport and contribution of advanced biofuels (Adapted from [7]).

their energy content) in terms of their contribution toward the 2030 target of 14% for renewable energy in transport. If the advanced fuels are used for marine or aviation transport, their counting will be multiplied by 1.2. Some feedstocks such as Used Cooking Oil (UCO) and animal fats will be double counted but have specific cap of 1.7%. For liquid biofuels, the default GHG emission values and calculation rules are provided in Annex V. The greenhouse gas saving thresholds for biofuels in transport are 50% for plants with an operation start date before October 2015, 60% after October 2015, and will be 65% from January 2021.

EU points out that biofuel feedstock typically comes from cropland that was previously used for other agricultural uses such as growing food or feed. Since this agricultural production is still necessary, it may lead to the extension of agriculture land into noncropland, possibly including areas with high carbon stock such as forests, wetlands, and peatlands. This process is known as indirect land use change (ILUC). As this may cause the release of CO_2 stored in trees and soil, ILUC risks negating the greenhouse gas savings that result from the use of biofuels. To address this issue, RED II introduced a new approach, setting limits on high ILUC-risk biofuels, bioliquids, and biomass fuels with a significant expansion in land with high carbon stock. These limits will affect the amount of these fuels that member states can count toward their national targets when calculating the overall national share of renewables and the share of renewables in transport. Member states will still be able to use fuels covered by these limits, but they will not be able to include them when calculating the fulfillment level of their renewable targets. These limits consist of a freeze at 2019 levels for the period of 2021–2023, which will gradually decrease from the end of 2023 to zero by 2030. The directive also introduces an exemption from these limits for biofuels, bioliquids, and biomass fuels certified as low ILUC risk.

Just for comparing transport with other energy-consuming sectors, bioenergy accounts for approximately 12.3% of the gross energy consumed in Europe, almost double the weight of other renewable energies that represent 6.8% [8] (**Figure 2**), and is mainly used for the generation of heat, followed by the production of electricity and, in a minor extent, for transport fuels.

The weight of bioenergy in the European energy mix has been increasing in recent years (**Figure 3**) and must continue to grow in order to meet the objective set by the European Union, so the availability of raw materials that meet environmental, economic, and social sustainability criteria is a concern [10]. This chapter summarizes the situation of the feedstocks used, their future availability, and potential

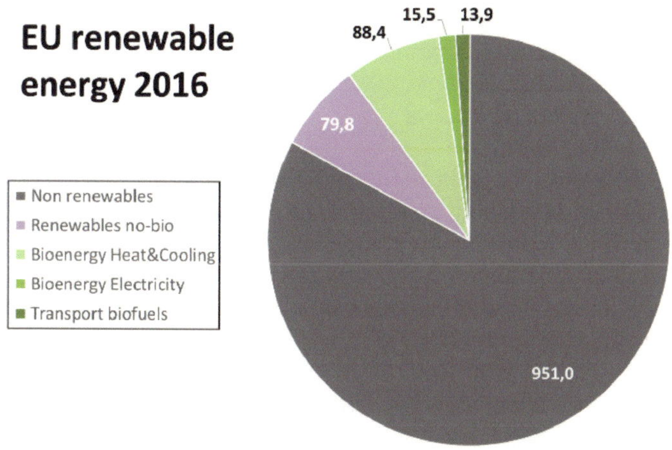

Figure 2.
Share of renewables in the EU's gross final energy consumption for 2016 (Mtoe/year) (Adapted from [8]).

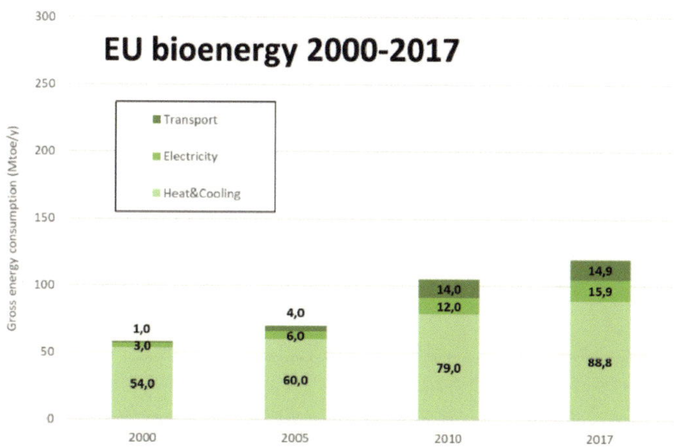

Figure 3.
Gross bioenergy consumption by markets 2000–2017 in EU28 (Mtoe/year) (Adapted from [9]).

supply according to the literature under different scenarios, with a special focus on aquatic biomass and some recommendations about the integration of the bioenergy sector with other bio-based industries to develop an integrated bioeconomy in Europe.

2. Advanced bioenergy and advanced biofuels

Biomass for electricity, heat, and cold (EHC) generation is used directly in solid form (wood, straw, etc.) or through intermediate energy carriers in solid form (pellets, chips, etc.), liquid form (bioliquids), or gas form (biogas). Although biogas can also be used as transport fuel, it undergoes a process of purification and reduction of impurities, then becoming biomethane. Solid fuels are sometimes treated to increase their energy density and storage stability through torrefaction or similar technologies, but minimizing changes needed in installed infrastructure.

In addition to applications for EHC, advanced bioenergy also includes advanced biofuels for transport. It is important to highlight that the objectives of advanced fuels for transport defined by RED II can not only be achieved through the use of advanced biofuels (produced from biomass) but also through other renewable fuels of nonbio origin, even from fossil carbon, if the energy needed to produce them is of renewable origin (**Figure 4**). All these options are clearly defined in the RED II [7] (**Table 1**).

2.1 Examples of advanced biofuels

Within the category of advanced biofuels for transport, a diverse range of products is included, depending on the feedstock used, the transformation process, and the application to which they are intended: road transport (diesel or gasoline engines), marine, or air transport (jet fuel). The most significant products are described below, some already available on a commercial scale and others with different degrees of development [11]:

- Hydrotreated vegetable oils (HVO)/hydroprocessed esters and fatty acids (HEFA) are chemically very similar to fossil diesel and kerosene. This fuel has

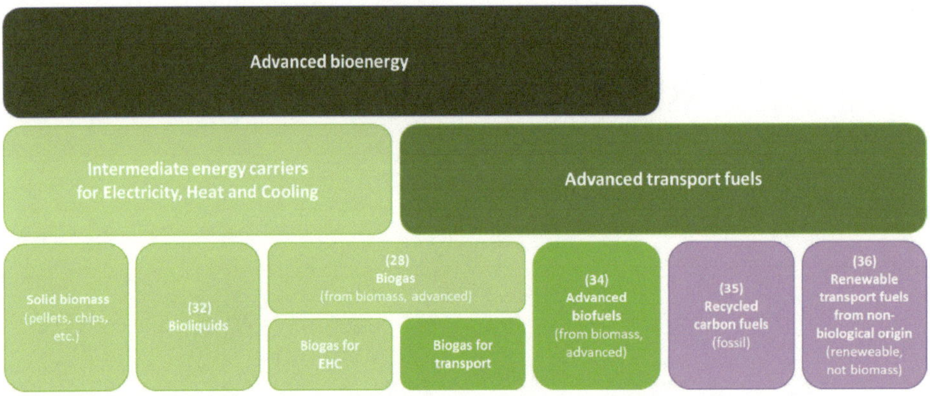

Figure 4.
Markets and products included in the advanced bioenergy definition (Adapted from [11]).

Some relevant definitions of RED II

(28) 'biogas' means gaseous fuels produced from biomass.

(32) 'bioliquids' means liquid fuel for energy purposes other than for transport, including electricity and heating and cooling, produced from biomass.

(33) 'biofuels' means liquid fuel for transport produced from biomass.

(34) 'advanced biofuels' means biofuels that are produced from the feedstock listed in Part A of Annex IX.

(35) recycled carbon fuels means liquid and gaseous fuels that are produced from liquid or solid waste streams of nonrenewable origin which are not suitable for material recovery in accordance with Article 4 of Directive 2008/98/EC, or from waste processing gas and exhaust gas of non-renewable origin which are produced as an unavoidable and unintentional consequence of the production process in industrial installations (GHG saving required for these fuels will be defined by the European Union before January 1, 2021.

(36) renewable liquid and gaseous transport fuels of non-biological origin means liquid or gaseous fuels which are used in the transport sector other than biofuels or biogas, the energy content of which is derived from renewable sources other than biomass.

Table 1.
Some relevant definitions of RED II [7].

clear advantages over the ester-type biodiesel fuels, such as lower NO_x emission, deposit formation, storage stability, not degrading the engine oil and better cold properties. HVOs are paraffinic hydrocarbons that are free of aromatics and heteroatoms and have excellent combustion properties (high cetane number). They are also approved to use as aviation fuels. Currently, HVOs are widely produced from vegetable oils, and the technological challenge for the industry is to produce HVOs from sustainable feedstocks. A by-product of HVO production is bio-propane produced from glycerin hydrogenation and that can be used as BioLPG (liquid petroleum gases, such as butane and propane).

- Cellulosic ethanol is the most common advanced biofuels for Otto engines. It is produced by hydrolysis and fermentation of lignocellulose from agricultural wastes (straw, corn stover, bagasse, etc.), forestry waste (branches, etc.), organic fraction of municipal solid waste (OFMSW), or energy crops. The end product is identical to bioethanol produced from sugarcane or starch crops, and the technological challenge for the industry is to lower the production costs of lignocellulosic sugars. Ethanol can be blended with gasoline directly or used to produce ethyl-tertbutyl-ether (ETBE) that can also be used as gasoline component.

- FT liquids/biomass-to-liquid (BTL) fuels are produced also via gasification (syngas) but in this case followed by conditioning and fuel synthesis via Fischer-Tropsch or similar alternative processes. BTL fuels are used in diesel engines and aviation engines. An alternative to conventional gasification is the use of high temperature plasma gasification that can be applied to a wider range of feedstocks.

- Biomethanol can also be produced from syngas via a thermochemical route similar to the Fischer-Tropsch process for BTL. It can be directly blended with gasoline or used to produce methyl-tertbutyl ether (MTBE) or dimethylether (BioDME) via catalytic dehydration. BioDME can also be synthesized directly from syngas. In standard conditions, DME is a gas and can be used as transport fuel in a similar way to LPG.

- Biobutanol is an alcohol more compatible with existing fuel infrastructures and engines than ethanol. It can be produced also from lignocellulosic sugars with fermentation techniques. Currently, this route is under development, and the yield of sugars to butanol is not good enough to be economically competitive to ethanol. The technological challenge is to use advanced biotechnology to improve the bacterial butanol-producing strains.

- Other liquid advanced biofuels that are being developed are synthetic paraffinic fuel/hydrocarbons via chemical or biotechnological catalysis of plant sugars. **These routes** offer great potential for converting sustainable sugars into drop-in fuels, molecules that have similar properties to fossil gasoline or diesel.

- Biosynthetic natural gas (BioSNG) is produced by gasification of biomass, followed by a gas conditioning step, SNG synthesis, and gas upgrading. BioSNG can be used in a similar way to biomethane (biogas) produced biologically through anaerobic digestion and that also needs an upgrading step to reduce impurities to be used as transport fuel.

- Biohydrogen can potentially be produced from biomass via various routes and used as transport fuel, but the yield is low, and renewable hydrogen will more probably be produced through electrolysis of photoelectrolysis of water using renewable electricity. An indirect route to produce biohydrogen is steam reforming of biogas or BioSNG.

3. Feedstocks considered for advanced bioenergy

According to the European regulation, advanced bioenergy is produced from feedstocks not used as food or feed, that does not compete with food or feed crops for resources such as soil and water, and that has a minimum greenhouse gas emission saving when compared to fossil fuels.

Some examples of feedstocks used for the EHC market to produce solid fuels, bioliquids, or biogas for this sector are [12]: agricultural (straw, hay, etc.) and

Feedstocks listed in Anex IX of RED II

Part A. Feedstocks for the production of biogas for transport and advanced biofuels, the contribution of which towards the minimum shares referred to in the first and fourth subparagraphs of Article 25(1) may be considered to be twice their energy content:

(a) Algae if cultivated on land in ponds or photobioreactors;
(b) Biomass fraction of mixed municipal waste, but not separated household waste subject to recycling targets under point (a) of Article 11(2) of Directive 2008/98/EC
(c) Biowaste as defined in point (4) of Article 3 of Directive 2008/98/EC from private households subject to separate collection as defined in point (11) of Article 3 of that Directive
(d) Biomass fraction of industrial waste not fit for use in the food or feed chain, including material from retail and wholesale and the agro-food and fish and aquaculture industry, and excluding feedstocks listed in part B of this Annex
(e) Straw
(f) Animal manure and sewage sludge
(g) Palm oil mill effluent and empty palm fruit bunches
(h) Tall oil pitch
(i) Crude glycerine
(j) Bagasse
(k) Grape marcs and wine lees
(l) Nut shells
(m) Husks
(n) Cobs cleaned of kernels of corn
(o) Biomass fraction of wastes and residues from forestry and forest-based industries, namely, bark, branches, precommercial thinnings, leaves, needles, tree tops, saw dust, cutter shavings, black liquor, brown liquor, fibre sludge, lignin and tall oil
(p) Other non-food cellulosic material
(q) Other ligno-cellulosic material except saw logs and veneer logs.

Part B. Feedstocks for the production of biofuels and biogas for transport, the contribution of which towards the minimum share established in the first subparagraph of Article 25(1) shall be limited and may be considered to be twice their energy content:

(a) Used cooking oil;
(b) Animal fats classified as categories 1 and 2 in accordance with Regulation (EC) No 1069/2009.

Table 2.
Feedstocks listed in the Annex IX of RED II for the production of biogas for transport and advanced biofuels [7].

animal residues (manure), forestry residues, natural conservation matter (urban maintenance of green areas, hay, and shrubs), roadside vegetation, and waste (urban, industrial, biodegradable municipal waste, selected waste from the food and wood industry). None of them is biomass that can be used for food and feed, and their GHG savings are high enough to be considered as advanced.

The situation in the transport sector is different, since most feedstocks used today for conventional biofuels are mostly cereals, sugar, and oil crops used in the food and feed sectors. For this reason, EU legislation clearly defines advanced biofuels by the feedstock used, listed in the Annex IX of the RED-II (**Table 2**) and the GHG savings produced, directly and considering also indirect effects (ILUC).

4. Feedstocks used today in Europe

According to the EurObserv'ER, total primary bioenergy consumption of the EU28 was 99.8 Mtoe of solid biomass in 2017 [12] and 17.0 Mtoe of biofuels in 2018 [13] (last data published). According to the European Bioenergy Day website [9], biomass mobilized in Europe to produce energy accounted for **144.1 Mtoe** in 2017, an amount that is near to overpass the European production of coal. Approximately 70% comes from forestry resources, 18% from agriculture, and 12% from wastes (**Figure 5**). Aquatic biomass (algae) used for bioenergy is currently irrelevant.

Approximately **two thirds** of the biomass consumed in Europe is solid biomass for the EHC market, being mostly forestry residues and agricultural by-products: wood from silviculture, waste wood, short rotation coppice, agricultural waste, and so on. Biogas and biofuels represent **11.7** and **1.4%**, respectively, of gross inland energy consumption of biomass. Finally, renewable municipal waste used directly for energy production is the fourth type of biomass for energy reaching **7.3%** in 2017 [14].

4.1 Feedstocks for the EHC market

Wood has always been the most popular source of biomass of energy in Europe. The residential sector is the main user of wood (**27%**) but is closely followed by the industrial use of wood chips (installations above 1 MW (**22%**)) and small-scale use of wood chips (**14%**). Pellet consumption in modern appliances is also growing, representing **6%** of total EU wood energy consumption. Historically, the European bioenergy sector developed in close synergy with other wood user industries to use

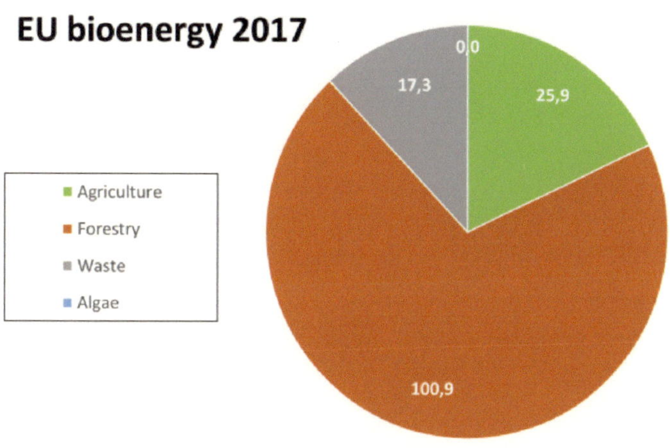

Figure 5.
Distribution of the biomass used for bioenergy in Europe in 2017 (Mtoe/year) (Adapted from [9]).

low value biomass such as thinnings, low-quality wood, tops and limbs, sawdust, or woodchips. In fact, bioenergy providers do not use any type of wood but mainly mobilize by-products of forest management operations and the wood industry.

In 2015, total amount of **245.2 million tons** of municipal waste was treated in Europe, of which **27% (67 million tons, mostly in Northern countries)** went to 492 waste-to-energy (incinerator) plants still remaining behind recycling (**30%**) and landfill (**24%**).

The European biogas sector is very diverse, depending on national priorities. In some countries, biogas production is seen as a waste management option, a renewable energy technology, or a combination of the two, countries have adapted their policies to favor certain feedstocks over others. Two countries represent the two ends of the scale: Germany and the UK. Germany produces **92%** of its biogas from agricultural crops and wastes, while in the UK, landfill and sewage sludge gas account for nearly 60% of the biogas production. Taking the EU28 big picture, field crops, manure, and agri-food industry waste represent around **75%** of the biomass used for biogas production, a share that **tripled** since 2010. Sewage sludge and landfills represent the **last 25%**. It is important to point out that although biogas is mainly used for the EHC market, its use as biomethane for the transport sector is increasing and should be taken into account in the future.

4.2 Feedstocks for the transport market

Today, the European biofuel sector is based on the use of bioethanol and biodiesel, which do not use the same feedstocks [15].

According to ePURE [16], the required feedstock for the 2018 production of ethanol (**5.81 Ml** of bioethanol) was mainly conventional not advanced (**11.11 Mt** of cereals, **2.07 Mt of sugar equivalent, mainly from sugar beets) and only 0.39 Mt of advanced (lignocellulosic and other listed in RED II Annex IX) (Figure 6).** This means that only about **2.9%** of the European cereal production and about **7.0%** of the sugar beet production are used to produce energy. Bioethanol production provides European farmers with **€6.6** billion income per year, and biofuels and food production can be mutually supportive, since **5.55** Mt. of co-products were produced in 2017 for nonenergy sectors, of which **4.20** Mt. was animal feed. Wheat is mainly used in northern Europe, while corn is used in Central Europe and Spain, and sugar beet users include France, Germany, and Belgium.

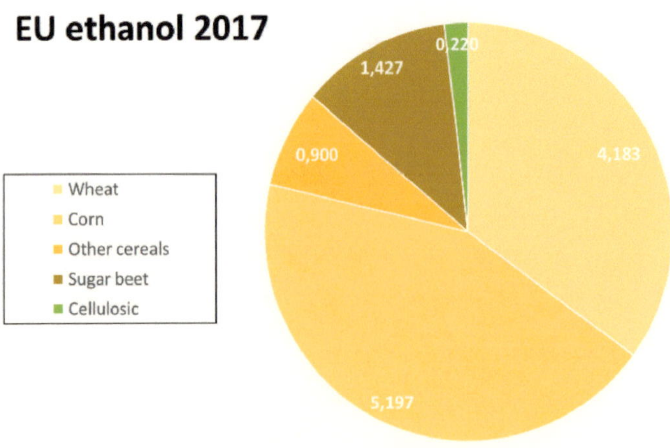

Figure 6.
Feedstock used for bioethanol production in Europe in 2017 (Mt/year, for sugar beet in Mt. sugar equivalent/ year) (Adapted from [16]).

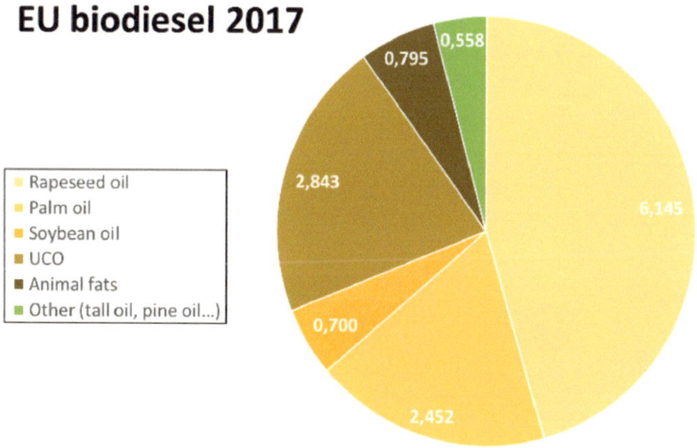

Figure 7.
Feedstock used for biodiesel production in Europe in 2017 (Mt/year) (Adapted from [15]).

Biodiesel's most used feedstock is rapeseed oil, accounting for **44%** of total production in 2017, according to Bioenergy Europe [15] (**Figure 7**). This is changing, mostly due to the higher use of palm oil and recycled vegetable oil/used cooking oil (UCO). In fact, UCO has become the second-most important feedstock in some countries such as the Netherlands, the UK, and Germany.

5. Future feedstock availability

The target of RED-II for advanced biofuels is to provide 3.5% of transport energy by 2030, multiple counting included. In the light of this ambitious target, and of the restrictions regarding raw materials that can be used, in recent years, different agencies and organizations have carried out studies to predict both the use and the expected and potential availability of raw materials in the short (2020), medium (2030), and long (2050) terms.

In a report for the European Commission, PriceWaterhouseCoopers and the EEV consortium [17] estimated the bioenergy demand per sector (**Figure 8**) and the feedstock supply to fulfill this demand (**Figure 9**) in 2030 in a scenario called Green-X EUCO27. The underlying policy concept of Green-X EUCO27 is to follow a least-cost approach for incentivizing the renewable requirements for 2030. The estimated demand of bioenergy in 2030 is 146.4 Mtoe/year (18.7 of them for transport), 36% more than in 2014. This demand will be fulfilled with 179.6 Mtoe/year of domestic biomass (plus 15.9 Mtoe/year of imported biomass and biofuels), 57% coming from forestry resources, 30% from agriculture, and 13% from waste and residues.

Regarding to the domestically available potential for biomass for energy, Bioenergy Europe based on literature review [14] calculated it to be between 169 and 737 Mtoe/year from 2050 onward (**Figure 10**). The middle range potential of 406 Mtoe, which is around 24% of the total energy consumption in EU28 in 2017, could be achieved by 2050—considering different constraints (e.g., costs). This means that, compared to the actual 144 Mtoe used in 2017, the potential gives enough room to almost triple the amount of bioenergy in the EU28 energy mix, with most resources coming from the agricultural sector.

Other review, coordinated by Ecorys for the European Commission [18], estimated the domestic biomass supply potential to be 562 Mt. dry mater/year in 2030 and 638 Mtdm/year in 2050 (under the reference scenario) and 700 Mtdm/year

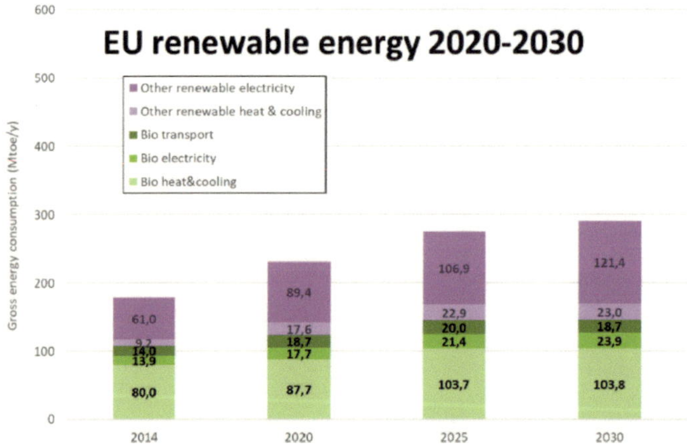

Figure 8.
Bioenergy and other renewable energy demand 2020–2030 in the Green-X EUCO27 scenario (Adapted from [17]).

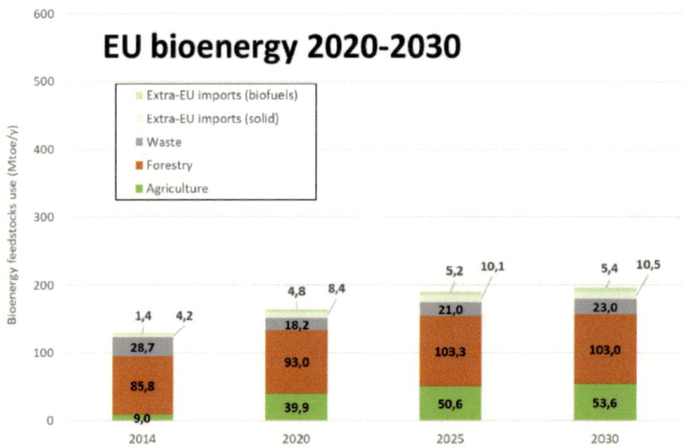

Figure 9.
Bioenergy feedstock supply 2020–2030 in the Green-X EUCO27 scenario (Adapted from [17]).

in 2030 and 1101 Mtdm/year in 2050 (under the high R&I scenario that boosts the potential of aquatic biomass) (**Figure 11**). These figures are slightly less optimistic than those of Bioenergy Europe, considering a conversion factor of 50% between tons of dry matter and tons of oil equivalent, but enough to triple the current bioenergy demand.

Some other references, such as ICCT, are not so optimistic about the biomass supply and express some doubts about the feasibility of decarbonizing all sectors together [19, 20].

In any case, the geographical distribution of the European potential biomass supply in 2030 is very different when it comes to agricultural, forestry, or waste origin, as described by the PWC study [17] and this something to consider when developing policies at a local level (**Figure 12**). Germany leads to the forestry potential, Spain to agricultural biomass, and the UK to biomass coming from wastes and residues.

Some other studies focus on the biomass availability predicted specifically for the advanced biofuel production, with the restricted criteria described in RED II. A study of the Arup URS consortium [21] provides a holistic analysis of the 28 feedstocks included in RED II Annex IX regarding the supply potentials, technology

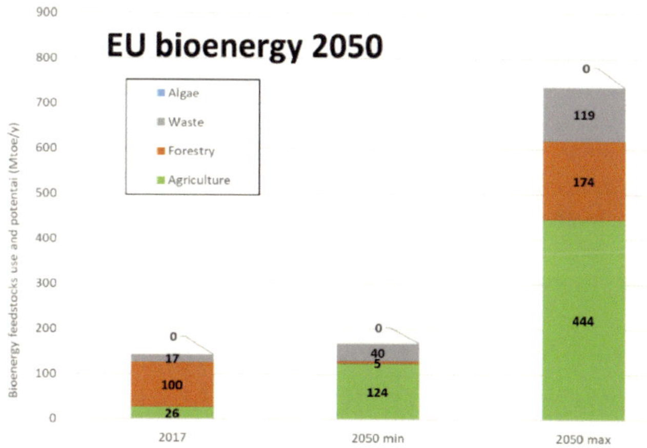

Figure 10.
Bioenergy feedstock potential 2050 in the EU28 MIN and MAX scenarios (Adapted from [14]).

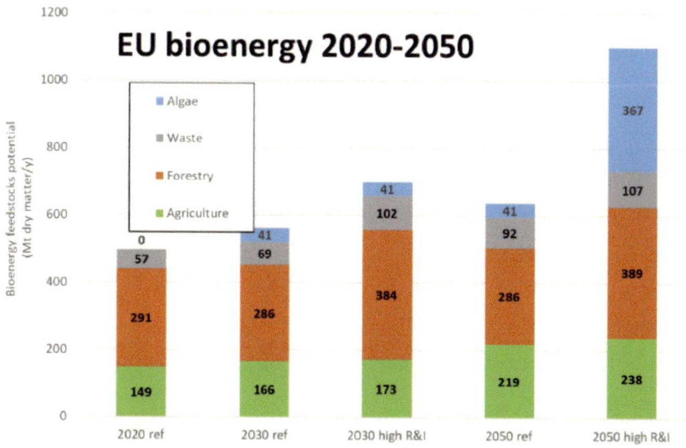

Figure 11.
Estimated potential biomass available 2020–2030–2050 in the EU in the reference and High R&I scenarios (Adapted from [18]).

Figure 12.
Distribution of the European biomass potential per country in 2030 for different feedstock origins: forestry (above), agriculture (middle), and wastes (below) (Adapted from [14]).

compatibility, economics, and sustainability (**Figure 13**). According to this study, the potential feedstock supply for advanced biofuels in Europe is around 130 Mtoe/year in the short term (2020). The main conclusions of the study can be summarized as follows:

- **Availability**: municipal and industrial wastes, straw, manures, forestry, and renewable electricity have the largest supply potentials. Wine residues, tall oil pitch, and crude glycerin are most limited. Energy crops, short rotation forestry, and algae will also be in short supply by 2020 but have potential in the long term.

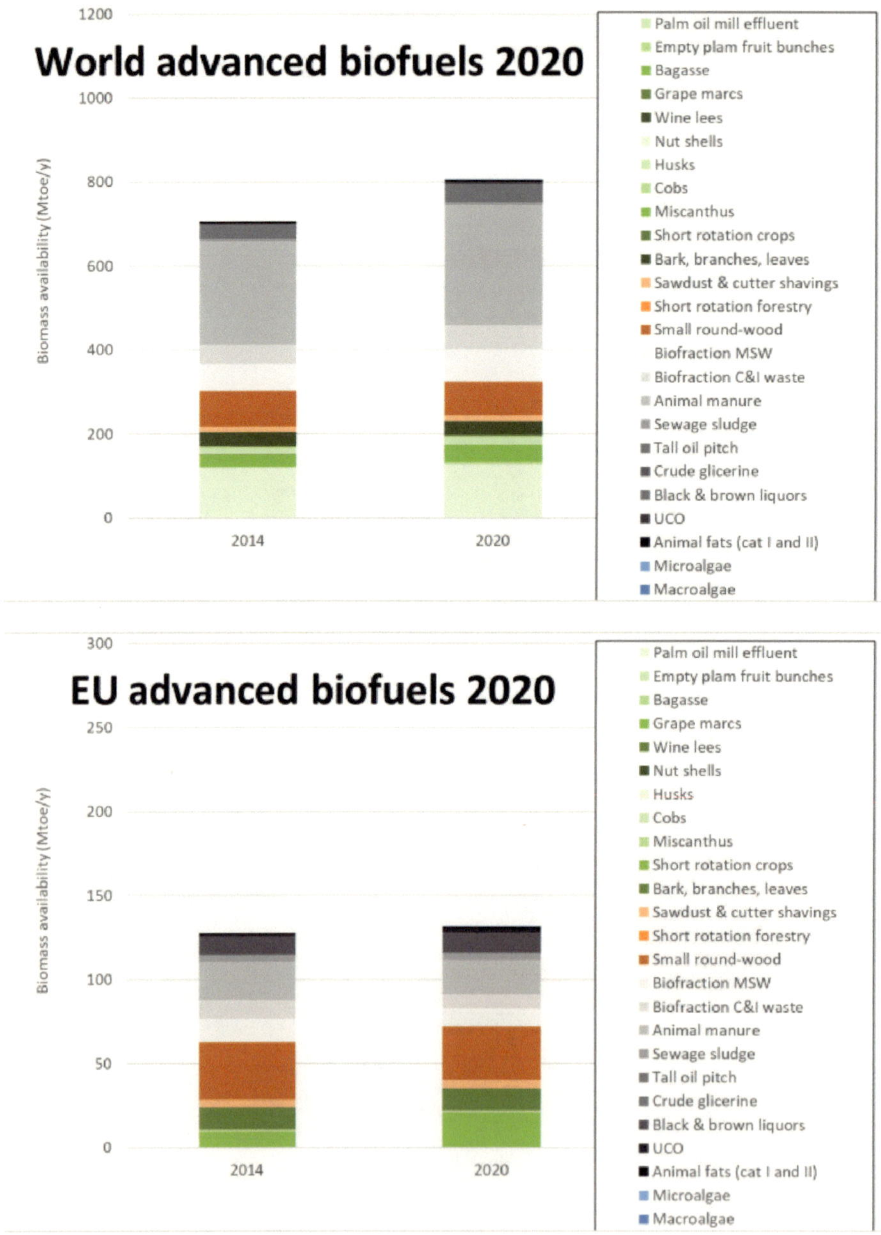

Figure 13.
Global 2020 feedstock supply potential for advanced biofuel production worldwide (above) and in the EU (below) (Adapted from [21]).

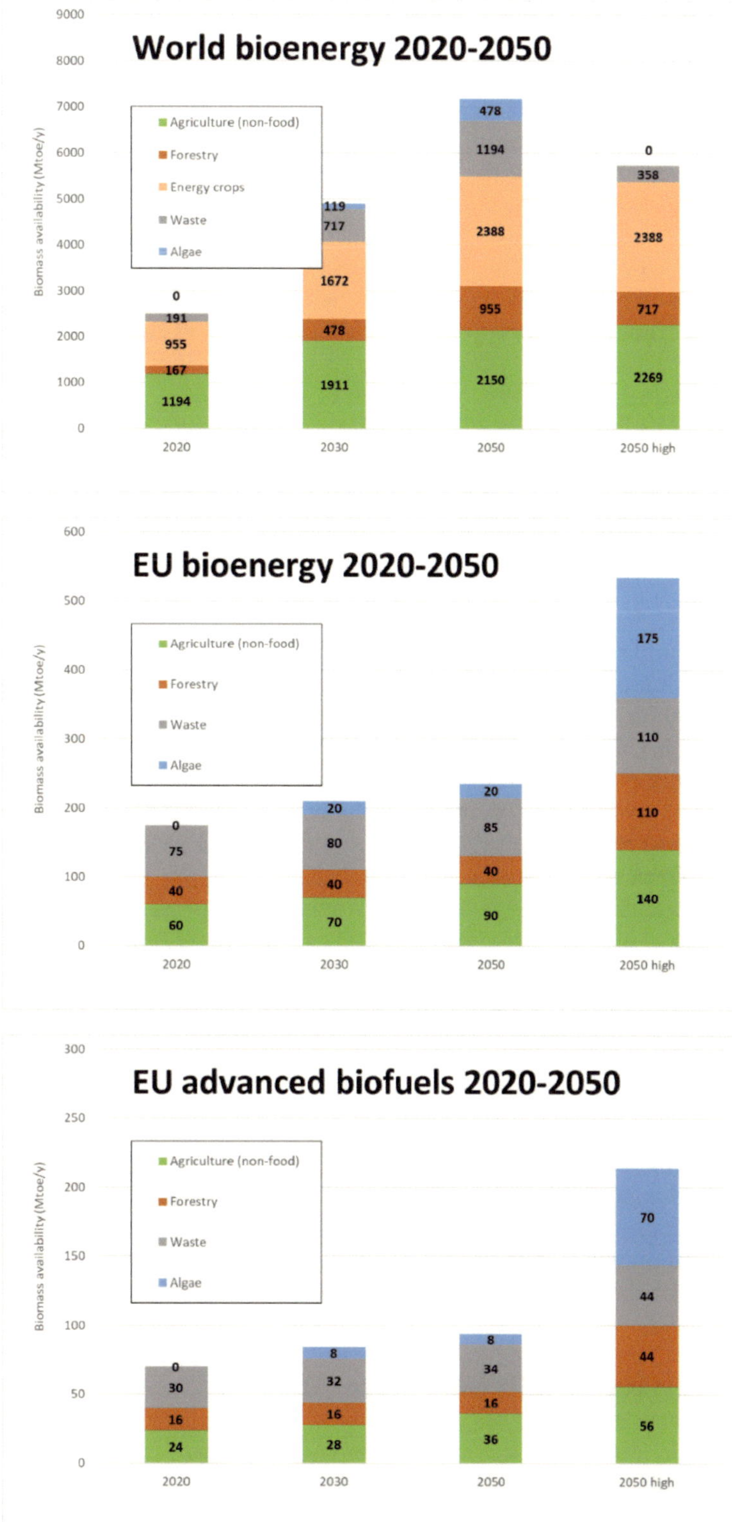

Figure 14.
Maximum biomass availability 2020–2030–2050 worldwide (above), in EU (middle), and in EU for transport (below) according to the IRENA (2020, 2030, and 2050) and IEA (2050 high) scenarios (Adapted from [22]).

- **Technology**: Many production technologies are still at pilot, demo or pre-commercial scale, including lignocellulosic ethanol and butanol, pyrolysis oil upgrading, gasification routes to alcohols, bio-Synthetic Natural Gas, Fischer-Tropsch diesel & jet, hydrothermal liquefaction (HTL), and renewable electrolysis. Some other technologies are commercially available but are only compatible with some of the Annex IX feedstocks.

- **GHG savings**: most feedstocks and routes are able to achieve GHG savings above 80%, but routes using MSW, C and I waste, bagasse, wine lees, algae, and waste carbon gases are more likely to fall into the 60–80% bracket, due to cultivation and transportation emissions and chemical and energy inputs.

Also with a focus on the transport sector, Concawe has also completed a literature review of the long-term availability of low-carbon feedstocks and fuels and the associated costs [22], describing scenarios for 2030 and 2050 in Europe and worldwide, based on IEA [23, 24], IRENA [25, 26], and other reports and reviews. When considering the availability of sustainable biomass in Europe, it should be noted that the whole of the bioenergy potential is estimated to grow from 175 Mtoe/year (2020) to approximately 350–535 Mtoe/year by 2050 (**Figure 14**).

According to the SGAB [27], the production of feedstock in Europe will be lower by 2050 and could range from 210 to 320 Mtoe/year (the majority coming from the waste sector). It is assumed that most of the biomass used in the EU economy will be produced within Europe (imports of sustainable solid biomass will be limited to 4–6% of the solid biomass used for bioenergy by 2050). For the transport sector, different sources estimate that the biomass contribution could range from 70 Mtoe/year (2020) to 140–210 Mtoe/year (2050). In terms of energy content, agricultural residues and wastes are expected to contribute the most, followed by forestry residues and algae.

World sustainable biomass availability is generally expected to increase continuously from a total of 2500 Mtoe/year by 2020 to 5700–7000 Mtoe/year by 2050 in the max scenario mainly based on agricultural residues and energy plants (>70%). The IEA 2050+ scenario forecasts a lower potential availability as defined by IRENA in their 2050 base scenario, with the main difference being the envisaged potential for algae. Indeed, the potential of algae is uncertain, and while several sources recognize its role in the 2050 scenario, other sources are more conservative and do not consider that there will be any relevant contribution before 2050. This point will be discussed later.

An interesting specific study highlights the potential of waste resources, "Waste—Europe's untapped resource" [28], and states that if all waste and residues were converted only to biofuels in the EU, 16% of road transport fuel could be provided in 2030 (technical potential of sustainably available feedstock from waste).

6. Algal biomass availability in Europe

Aquatic biomass, mainly macro and microalgae, is one of the most controversial feedstocks for advanced biofuels. It has attracted great interest in the recent years, as it does not compete with food crops for land use, does not need freshwater, and can absorb a great amount of carbon dioxide, and its productivity per hectare is much higher than any terrestrial crop [29, 30]. Some studies are optimistic about its weight in the future global biomass availability, but many others think that its techno-economic maturity is and will be far from being competitive for the bioenergy sector and that economics call for higher-value products, such as cosmetics, pigments, food supplements, proteins, and additives.

As with many technologies at currently low TRL, a lot of uncertainty surrounds the potential evolution of production costs for aquatic biomass. A qualitative analysis carried out [18] revealed that the technical potential of aquatic biomass produced domestically in the EU is significant but is quite limited due to the expected production costs. However, the level of uncertainty behind those figures is important; hence, a sensitivity analysis was conducted in order to assess the contribution of algae in the bioenergy mix depending on the production costs (**Figure 15**).

Another detailed study [31] based on a GIS model showed that algal biomass potential is limited by the limited availability of marginal land in densely populated Europe and by the slope of many of these areas or their status as protected areas. A general result was that about 50 Mt. dry matter microalgal biomass could be produced annually in Europe. This figure is in line with the 2030–2050 projections of IRENA (20 Mtoe/year) but far from the very optimistic IEA 2050+ scenario (175 Mtoe/year). The by far largest part would come from Spain (34 Mt/year), but countries such as Sweden, Italy, and Portugal also show considerable potential (**Figure 16**).

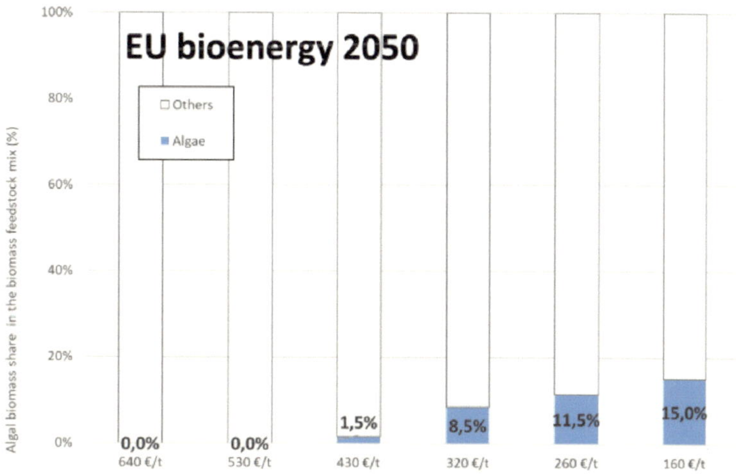

Figure 15.
Share of aquatic biomass penetration in the biomass feedstock mix 2050 under different productions costs for microalgae (Adapted from [18]).

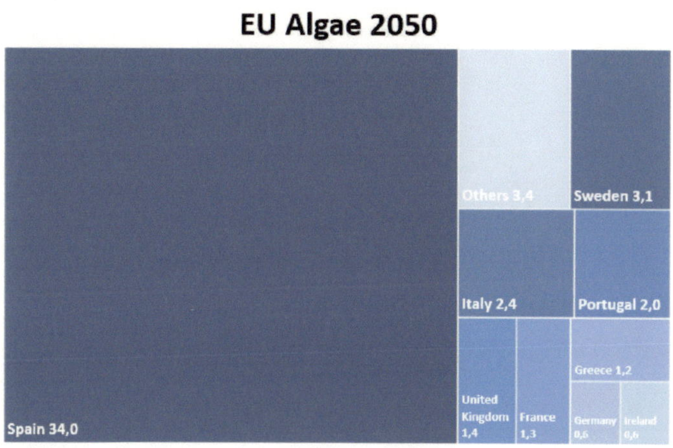

Figure 16.
Maximum microalgae biomass resource potential per country in EU (Mt dry biomass per year) (Adapted from [31]).

7. Developing an integrated bioeconomy in Europe

A recent study about the distribution of biorefineries in Europe [32] shows a total of 803 biorefineries identified. Only 177 of them (22%) are biorefineries in which integrated production of bio-based products (chemicals and/or composites) and bio-based energy (biofuels and/or other types of energy from biomass) is taking place, and that thus reflects the strictest definition of biorefinery. A certain degree of correspondence between the location of biorefineries and the locations of ports and of chemical clusters in the EU can be observed. The highest density of biorefineries is in Belgium, the Netherlands, and some highly industrialized regions of Germany, France, and Italy. The lower number of biorefineries in the Eastern part of the EU demonstrates an untapped potential. Worldwide mapping on the advanced biofuel production facilities is done by IEA Bioenergy Task 39 in its online database [33].

The sustainable supply of biomass for the energy sector passes in the medium term through the organization, coordination, and integration with the different sectors interested in available biomass resources. This includes both the productive sectors (agricultural, forestry, and waste management) and users (food, bio-based materials, and chemicals), so that supply chains and value chains are optimized, leading to an integrated economic sector based on the bioeconomy. The balanced distribution of available resources and the criteria or tools for decision making have been reviewed in different studies by FAO and others [34–36]. The EU is also responding to such an ambitious challenge by funding different projects aiming to develop, scale up, and deploy new technologies and value chains from an integrated approach [37]. Some of these projects are listed in **Table 3**, and the key R&D lines to boost bioeconomy are listed in **Table 4**.

Project acronym	Feedstock	Coordinating country	Years
PROMINENT	Cereal processing side streams	Finland	2015-18
VALCHEM	Woody feedstock	Finland	2015-19
US4GREENCHEM	Lignocellulosic feedstock	Germany	2015-19
FIRST2RUN	Cardoon from marginal lands	Italy	2015-19
PULP2VALUE	Sugarbeet pulp	Netherlands	2015-19
SMARTLI	Karft lignin lignosulphonates	Finland	2015-19
STAR4BBI	Lignocell. agri and forest residues	Netherlands	2016-19
BIORESCUE	Wheat straw and agricultural waste	Spain	2016-20
AGRIMAX	Agri and food waste	Spain	2016-20
GREENSOIREX	Lignocellulosic residues	Netherlands	2016-20
GRRENPROTEIN	Veg residues from salad processing	Netherlands	2016-20
ZELCOR	Lignocellulosic residues	France	2016-20
FUNGUSCHAIN	Mushrooms farming residues	Netherlands	2016-20
LIBBIO	Andes lupin from marginal lands	Iceland	2016-20
BIOSKOH	Lignocellulosic feedstock	Italy	2016-22
LIGNOFLAG	Straw	Germany	2017-20
BARBARA	Agri and food waste	Spain	2017-20
SYLEED	Wood residues	France	2017-20
POLYBIOSKIN	Food waste	Spain	2017-20
OPTISOCHEM	Residual wheat straw	France	2017-21
4REFINERY	Biomass, wastes	Norway	2017-21
DENDROMASS4EUROPE	Dendromass on marginal lands	Germany	2017-22
GRACE	Miscanthus, hemp on marginal lands	Germany	2017-22
AGRICHEMWEY	Byproducts from dairy processing	Ireland	2018-21

Table 3.
Some projects recently funded by EU to develop a bioeconomy based on advanced biomass feedstocks (Adapted from [37]).

Biomass from agriculture

- Improving biomass cultivation (cropping)
  - Food and energy crops breeding
  - Agricultural practices improving
  - Crop rotation and inter-cropping application
  - Agroforestry and short-rotation crops development
  - Marginal land for energy crops development
- Improving biomass harvesting
- Improving biomass pretreatment and densification
- Improving biomass supply chain
  - Biomass mobilisation optimisation
  - Agricultural biomass logistics optimisation
  - Agricultural biomass supply chain optimisation
  - Technology transfer enhancement

Biomass from forestry

- Improving forest biomass production
  - Breeding of genetically improved plant material
  - Fertilisation improvement
  - Silviculture impriving
- Improving forest biomass production
  - Breeding of genetically improved plant material
  - Fertilisation improvement
  - Silviculture impriving
- Improving biomass suply chain
  - Biomass mobilisation optimisation
  - Forestry biomass logistics optimisation
  - Forestry biomass suplly chain optimisation

Biomass from waste

- Optimising supply chain
  - Source-separated biowaste collection and use
  - Mechanically-separated biowaste collection and use
  - Landfilled biowaste use
  - Use of UCO, woodwaste, vegetal waste, paper and cardboard waste, textile waste, sewage sludge

Aquatic (algae) biomass

- Improving microalgae cultivation systems and productivity
  - Cultivation in open ponds
  - Cultivation in photoreactors
- Improving macroalgae cultivation systems and productivity
  - Wild cultivation
  - Aquafarms (mariculture)
- Improving micro and macroalgae harvesting, and dewatering
- Improving microalgae lipids extraction
- Improving microalgae lipid-to-fuel conversion technologies
- Improving micro and macroalgae GHG balance

Table 4.
Priority R&D lines to develop a sustainable biomass supply for and integrated bioeconomy in Europe (Adapted from [18, 24, 26]).

8. Conclusions

Advanced biofuels can be a key factor to a sustainable energy supply for transport, contributing to energy security, fight against climate change, and development of rural areas. However, Europe has to overcome some barriers in order to develop their full potential.

First, a stable demand is needed to establish a market and boost development, with production levels sufficient to achieve economies of scale. Current reasons that slow uptake are high barriers to entry, including long investment cycles, the capital-intensive nature, high fuel certification standards, and high production costs compared to fossil fuels and conventional biofuels. To overcome some of these barriers, the key points to consider can be summarized as follows:

- guarantying sustainable feedstock availability, minimizing supply chain risks, and mobilizing currently unexploited sustainable waste, biomass, and other resources;

- facilitating agricultural and forestry development of energy crops with higher productivity with low chemicals and energy input and using marginal land that does not compete directly with or displace land used for food crops;

- supporting the development of low-TRL technologies to produce advanced biofuels, increasing their efficiency, and of high-TRL technologies to deploy, reducing costs, and complying with GHG emissions goals;

- more specifically, supporting the development of technologies necessary to adapt advanced feedstocks into existing industrial processes (pretreatments);

- integrating the biofuels industry in a European bioeconomy system to take advantage of the opportunities both economies of scale and the use of existing infrastructure;

- recognizing of the role of advanced biofuels in transport, through a holistic approach across the whole well-to-wheels or even life-cycle value chain with a particular emphasis on the application of the principles of a circular economy.

Conflict of interest

The authors declare no conflict of interest.

DOI: http://dx.doi.org/10.5772/intechopen.91006

Author details

Enrique Espí*, Íñigo Ribas, Carlos Díaz and Óscar Sastrón
Repsol SA, Madrid, Spain

*Address all correspondence to: eespig@repsol.com

References

[1] European Commission. A Clean Planet for all. A European Strategic Long-Term Vision for a Prosperous, Modern, Competitive and Climate Neutral Economy. In-Depth Analysis in Support of the Commission Communication. Brussels: COM; 2018

[2] European Commission. Sustainability Criteria. Proposal for Updated Sustainability Criteria for Biofuels, Bioliquids and Biomass Fuels. EC Website. 2019. Available from: https://ec.europa.eu/energy/en/topics/renewable-energy/biofuels/sustainability-criteria

[3] European Union. Directive 2009/28/EC of the European Parliament and of the Council of 23 April 2009. Brussels: Office of the European Union; 2009

[4] European Union. Directive 2009/30/EC of the European Parliament and of the Council of 23 April 2009. Brussels: Office of European Union; 2009

[5] European Union. Directive (EU) 2015/1513 of the European Parliament and of the Council of 9 September 2015 Amending Directive 98/70/EC Relating to the Quality of Petrol and Diesel Fuels and Amending Directive 2009/28/EC on the Promotion of the Use of Energy from Renewable Sources. Brussels: Office of the European Union; 2015

[6] European Union. Commission Delegated Regulation (EU) 2019/807 of 13 March 2019 Supplementing Directive (EU) 2018/2001 of the European Parliament and of the Council as Regards the Determination of High Indirect Land-Use Change-Risk Feedstock for which a Significant Expansion of the Production Area into Land with High Carbon Stock is Observed and the Certification of low Indirect Land-Use Change-Risk Biofuels, Bioliquids and Biomass Fuels. Brussels: Office of the European Union; 2019

[7] European Union. Directive 2018/2001 of the European Parliament and of the Council of 11 December 2018. Brussels: Office of the European Union; 2018

[8] European Commission. The European Commission's Knowledge Center for Bioeconomy. Brief on Biomass for Energy in the European Union. Brussels: European Commission; 2019. Available from: https://ec.europa.eu/jrc/en/publication/brief-biomass-energy-european-union

[9] European Bioenergy Day. What Are the Biomass Sources Consumed in the EU-28? Brussels. 2019. Available from: http://www.europeanbioenergyday.eu/bioenergy-facts/bioenergy-in-europe/what-are-the-volumes-of-biomass-used-in-the-eu28-to-produce-energy/

[10] AEBIOM. Statistical Report 2017. Brussels: Bioenergy Europe; 2017

[11] ETIP Bioenergy. Which Different Types of Advanced Bioenergy Are There? ETIP Bioenergy Website. Available from: http://www.etipbioenergy.eu/advanced-biofuels-overview

[12] EurObservER. Solid Biomass Barometer 2018. Paris: EurObserv'ER; 2018

[13] EurObserER. Biofuels Barometer 2019. Paris: EurObserv'ER; 2019

[14] Calderón C et al. Bioenergy Europe Statistical Report 2019. Biofuels from Transport. Brussels: Bioenergy Europe; 2019. Available from: https://bioenergyeurope.org/statistical-report.html

[15] Calderón C et al. Bioenergy Europe Statistical Report 2019. Biomass Supply. Brussels: Bioenergy Europe; 2019. Available from: https://bioenergyeurope.org/statistical-report.html

[16] ePURE. European Renewable Ethanol. Key Figures 2018. Brussels: ePURE; 2019. Available from: https://www.epure.org/resources/statistics/

[17] Hoefnagels R et al. Biomass Supply Potentials for the EU and Biomass Demand from the Material Sector by 2030. Sustainable and Optimal Use of Biomass for Energy in the EU Beyond 2020. Brussels: European Commission; 2017

[18] European Union. Research and Innovation Perspective of the Mid- and Long-Term Potential of Advanced Biofuels in Europe. Luxembourg: European Union; 2017

[19] Searle S. Bioenergy Can Solve Some of Our Climate Problems, But Not All of Them at Once. ICCT Website. Available from: https://theicct.org/blog/staff/bioenergy-solve-some-climate-problems-not-all-once

[20] Pavlenko N et al. Beyond the Biofrontier: Balancing Competing Uses for the Biomass Resource. Washington: ICCT; 2016. Available from: https://theicct.org/publications/beyond-biofrontier-balancing-competing-uses-biomass-resource

[21] Arup URS Consortium. Advanced Biofuels Feedstocks—An Assessment of Sustainability. London: UK Department of Transport; 2014

[22] Concawe. A look into the maximum potential availability and demand for low-carbon feedstocks/fuels in Europe (2020-2050). Concawe Review. 2019;**27**:4-20

[23] IEA Bioenergy. Annual Report 2018. Paris: IEA; 2019

[24] IEA. Technology Roadmap—Delivering Sustainable Bioenergy. Paris: OECD/IEA; 2017

[25] IRENA. Remap 2030. A Renewable Energy Roadmap 2030. Abu Dhabi: IRENA; 2014

[26] IRENA. Innovation Outlook. Advanced Liquid Biofuels. Abu Dhabi: IRENA; 2016

[27] European Commission. Building up the Future. Sustainable Transport Forum. Subgroup on Advanced Biofuels. Final Report. Luxembourg: European Commission; 2017

[28] Harrison P, editor. Wasted. Europe's Untapped Resource. An Assessment of Advanced Biofuels from Wastes and Resources. Washington: ICCT; 2014. Available from: https://theicct.org/publications/wasted-europes-untapped-resource

[29] FAO. Algae-Based Biofuels. Applications and Co-Products. Rome: FAO; 2010

[30] Rocca S et al. Biofuels from Algae: Technology Options, Energy Balance and GHG Emissions: Insights from a Literature Review. JRC. Brussels: European Commission; 2015

[31] Skarta J. Microalgae biomass potential in Europe. Land availability as a key issue. Technikfolgenabschätzung – Theorie und Praxis. 2012;**21**(1):72-79

[32] Parisi C. Research Brief: Biorefineries Distribution in the EU. European Commission—Joint Research Centre, Brussels, 2018. BEFS. Bioenergy and Food Security: The BEFS Analytical Framework. Environment and Natural Resources Management Series No. 16. Rome: FAO; 2010

[33] IEA Bioenergy. Database on Facilities for the Production of

Advanced Liquid and Gaseous Biofuels for Transport. IEA Website. 2019. Available from: https://demoplants.bioenergy2020.eu/

[34] FAO-UNEP. A Decision Support Tool for Sustainable Bioenergy. An Overview. Rome: UN Energy Publication; 2010

[35] Gomez San Juan M, Bogdanski A, Dubois O. Towards Sustainable Bioeconomy—Lessons Learned from Case Studies. Rome: FAO; 2019

[36] UFOP. UFOP Report on Global Market Supply 2017/2018 European and World Demand for Biomass for the Purpose of Biofuel Production in Relation to Supply in the Food and Feedstuff Markets. Berlin: UFOP; 2019

[37] Ahorsu R et al. Significance and challenges of biomass as a suitable feedstock for bioenergy and biochemical production: A review. Energies. 2018;**11**:3366-3385

Chapter 4

Valorization of Microalgae and Energy Resources

Cynthia V. González-López, Francisco García-Cuadra, Natalia Jawiarczyk, José M. Fernández-Sevilla and Francisco G. Acién-Fernández

Abstract

Microalgae biotechnology has grown very rapidly in the last few decades due to the multiple applications that these microorganisms have from pharmaceuticals and cosmetics to foods/feeds and biofuels. One of the main challenges in expanding this industry is to enlarge the single use of the biomass produced in addition to reducing the high biomass production cost of the current technologies. To overcome this bottleneck, the development of microalgae-based biorefineries has been proposed. The issue is to obtain as many bioproducts as possible from the cultivated biomass, including biofuels. Consequently, biodiesel production (from the lipid fraction), bioethanol (from carbohydrate fraction), and biogas or bio-oil (from the whole biomass) have been posited. In this book chapter, we review the current state of the art in the production of sustainable biofuels from microalgae and analyze the potential of microalgae to contribute to the biofuel sector.

Keywords: microalgae, biodiesel, bioethanol, biogas, biorefinery

1. Introduction

The policy trends regarding fuel consumption and greenhouse gas emissions are heading progressively towards an increase in renewable alternatives over the coming decades. There has been a significant rise in the production of solar, wind, marine, hydro, geothermal and biomass energy [1]. With regard to the transport sector, apart from electrification using renewable electricity, biofuels are the only sustainable alternative for reducing greenhouse gas (GHG) emissions. The current share of renewables in the transport sector is almost 3% as a result of using biofuels [2]. In 2014, global liquid biofuel production was 126 million m^3 of which 78 were bioethanol, 32 were biodiesel and 16 were advanced biofuels [2]. Focusing on the biomass, this can be converted into first, second or third-generation biofuels depending on the source of the organic material. First-generation biofuels are produced from food crops or vegetables. A serious debate still exists about their use as a source of biofuels given the land-use competition with food. Because of this, second-generation biofuels were developed to overcome the first-generation limitations. These are produced from non-food crops such as wood, organic waste, food crop waste and specific biomass crops such as jatropha or jojoba. However, the production capacity is low, so third generation biofuels were proposed to improve process sustainability.

Microalgae are the feedstock for third-generation biofuels. They possess several advantages over higher plants especially their rapid growth rate and photosynthetic yield [3]. Furthermore, they can be cultivated on non-arable lands and using water sources that are not intended for human consumption [4]. Their high metabolic plasticity is of great importance, as it allows the biomass composition to be modulated in response to culture conditions, thus the production of the metabolite of interest can be augmented [5]. However, the large-scale commercialization of biofuels from microalgae looks unlikely, mainly because of its negative economic balance resulting from the high production and processing cost of the biomass [6]. Nevertheless, an approach using the biorefinery concept would allow numerous compounds of interest to be obtained from the microalgae instead of just a single biofuel. The biomass would be fractionated into its main components, obtaining various bioproducts from each of them. The cultivation of microalgae must also be coupled with a reduction in greenhouse gas emissions by utilizing exhaust combustion gases as well as wastewater treatment services [7]. In this way, the generation of biofuels from microalgae would add to the contribution made by other types of renewable sources.

The 2011 biofuels roadmap used in the transportation sector [8] stated that up to 27% of worldwide transportation fuel could be supplied from biofuels by 2050. The European Union (EU) has established a 7% limit on the consumption of biofuels produced from agricultural crops in favor of advanced biofuels obtained from other feedstocks, which includes microalgae [9]. Within the international framework, several projects focusing on biofuels from microalgae production have been promoted over recent years [8]. In the United States of America (USA), the greatest producer of biomass-based biofuels, the National Renewable Energy Laboratory (NREL) resumed the R&D program for microalgae biofuels in 2006, and the National Alliance for Algal Biofuels and Bioproducts (NAABB) received 44 M$ of funding in 2009 for three integrated biorefinery demonstration plants—Solazyme, Algenol and Sapphire—with a combined funding total of almost 97 M$. In addition, several consortia were created, such as the Algal Biomass Organization (ABO), the Consortium for Algal Biofuels Commercialization (CAB-Comm), the Sustainable Algal Biofuels Consortium (SABC), the Cornell Marine Algal Biofuels Consortium, the Algae Testbed Public Private Partnership (ATP3) and the Regional Algal Feedstock Testbed (RAFT).

In the EU, the European Commission's (EC) 7th Framework Program (FP7) supported the development of biofuels from microalgae projects such as Sunbiopath, Algadisk and Aquafuels (2009) with the objective of selecting microalgae strains that are valuable for biofuel production, optimizing culture conditions and developing techno-economic reports for microalgae biorefineries. As a result, the European Algae Biomass Association (EABA) was created. In the United Kingdom, the Algae Biofuels Challenge was launched in the Advanced Bioenergy Accelerators strategy framework and promoted by The Carbon Trust. They are also developing the InteSusAl Project, having already built a 1 Ha pilot plant in Olhão (Portugal). In Germany, the BioEnergie 2021 program started in 2004 and led to the establishment of the European Algae Biomass Association (EABA) in 2009. Portugal obtained funding for the BioFAT Project, having built a 0.5 Ha pilot plant in Pataias (Portugal) and another in Camporosso (Italy) includes a projected construction of a 10 Ha facility. Spain has also shown great interest in biofuels and funded several projects using microalgae as the feedstock, such as CENIT CO2, MENOS CO2, Plan E, NOVARE and ALGAPLANE, which have also led to the construction of 2 pilot plants in Almería (Spain). Nowadays, All-Gas and SABANA Projects are currently underway. The objective of All-Gas is to develop a 10 Ha microalgae facility for biofuel production. SABANA is funded by the European Union's Horizon 2020

Research and Innovation program and is overseen by the University of Almería (Spain). SABANA aims to develop a large-scale integrated microalgae-based biorefinery based on a zero-waste process concept that is environmentally and economically sustainable. The EU is also presently funding several other projects in this field: AlgaeBioGas, which uses microalgae for the treatment of biogas digestate; DEMA, which produces ethanol; D-Factory, whose objective is to demonstrate the feasibility of *Dunaliella* cultivation at the large-scale (100 Ha); EnAlgae, which operates several pilot plants; and Fuel4ME, which produces biofuels at the pilot scale.

2. Large-scale production of microalgae biomass

Microalgae production can be accomplished in open systems such as open ponds or raceways, or in closed systems such as tubular photobioreactors or flat panels. Open systems are well established for commercial microalgae production, as they are the cheapest to operate and offer lower energy capture efficiency. For these reasons, facilities of up to 200 ha have recently been constructed in China although most conventional systems average up to 20 Ha. When using raceway reactors, the facilities comprise 5000 m^3 units operated in batch or semi-continuous mode, which achieve biomass productivities from 20 to 60 $T·Ha^{-1}·year^{-1}$. When high value products are the target, then closed systems (PBRs) are chosen. Furthermore, PBRs provide higher volumetric biomass productivities, which involve lower harvesting-associated costs. In this case, much smaller facilities are in operation worldwide, ranging from 1 to 10 Ha; these are typically composed of multiple reactors, each up to 20 m^3. When using closed reactors, biomass productivities vary between 40 and 80 $T·Ha^{-1}·year^{-1}$ although the most significant advantage is the high quality of the biomass produced. Whichever the type, the overall biomass productivity from any microalgae-related system depends on the production technology used, the geographic location and the culture conditions [10–12]. Light availability is the key factor for biomass generation, while for large-scale microalgae production, large amounts of nutrients are also required. Consequently, the only way to make the process feasible is to provide nutrients from waste.

To be sustainable, the energy balance in microalgae-related systems must be positive, so the Net Energy Ratio (NER) of the process must be considered. This represents the energy gained divided by the energy consumed. A value above 1.0 is required to make the process feasible. It has been shown that values of 1.01 are achievable with flat panels and 1.40 with raceways, whereas with tubular photobioreactors, a value of only 0.21 has been achieved [13, 14]. In microalgae-related systems, the energy consumption of the biomass processing step is very relevant. Therefore, depending on the technology used for this step, the energy consumption can be really high, making the use of biological processes more recommendable. In the next section, we discuss the various treatment possibilities.

3. Production of biofuels from microalgae

3.1 Bioethanol production

Bioethanol is the biofuel produced in the greatest quantity worldwide and it is used mainly in the transport sector [15] as a gasoline additive. The biggest producer in the world is the USA at around 60.9 million m^3 (56%), followed by Brazil at around 30.1 (28%), then Europe at 5.4 (5%) [16]. Bioethanol can be used as a substitute for petrol, as a blend component, or as a feedstock to produce Ethyl Tertiary Butyl Ether (ETBE), which is an additive that improves the combustion characteristics of petrol. ETBE is produced from ethanol and isobutylene in a catalytic reaction and it represents 60% of ethanol consumption; this is expected to remain constant until 2030 [17].

Nowadays, bioethanol is produced from feedstocks such as sugar crops (mainly sugarcane, sugar beet or sweet sorghum), starch crops (mainly corn; wheat is the main crop in Europe) or lignocellulosic biomass (agricultural and forestry residues and energy crops). Using microalgae as a source for third-generation bioethanol production has been proposed because of their various advantages over higher plants [18]. It has been suggested that microalgae could produce up to 140 $m^3 \cdot Ha^{-1} \cdot year^{-1}$ of ethanol compared to values below 8.0 $m^3 \cdot Ha^{-1} \cdot year^{-1}$ for sugarcane or sugar beet [19]. If a low biomass productivity of 20 $g \cdot m^{-2}$ is assumed, an annual yield of $73 \cdot 10^3$ $kg \cdot Ha^{-1}$ would be achieved, which is comparable to that of sugarcane crops (reported at $70–77 \cdot 10^3$ kg Ha^{-1}, equivalent to an ethanol production of 5–7 $m^3 \cdot Ha^{-1} \cdot year^{-1}$) [20]. Microalgae biomass contains polysaccharides (cellulose or starch) and other complex carbohydrates composed of monosaccharides like glucose, galactose, mannose, xylose, ribose, arabinose and other sugars [20]. Some microalgae and cyanobacteria are even able to excrete exopolysaccharides into the culture broth [21]. Several species have been proposed for bioethanol production because of their high carbohydrate content [22] although culture conditions can greatly modify the biochemical composition of any selected strain. This makes culture optimization and process control a key issue to tackle.

The process of bioethanol generation from previously harvested microalgae begins with releasing the polysaccharides from the biomass and converting them into fermentable sugars [18]. Thus, the first step is to break the cell wall and perform physical, enzymatic or chemical hydrolysis. Of these, acid hydrolysis is the most widely used as it provides a good conversion yield at a lower cost [23, 24]. The use of sulfuric, hydrochloric, or nitric acid at temperatures of 120–140°C for 15–30 min allows saccharification and fermentation yields higher than 80% [25] to be obtained. If enzymatic hydrolysis is chosen, amylases, cellulases, and/or pectinanes are available depending on the types of carbohydrates coming from the biomass, achieving similar ethanol yields.

Hydrolysis is followed by the alcoholic fermentation of the sugars to ethanol. The microorganism used to carry out the fermentation should be selected based on the specific sugars released from the microalgal biomass. Traditionally, the most widely used is *Saccharomyces cerevisiae* although other yeasts such as *Pichia stipitis* could be used, or bacteria such as *Zymomonas mobilis* and even recombinant bacteria (e.g. modified *Escherichia coli*) that are tolerant to high ethanol concentrations [26]. It is important to adjust the pH to make it optimal for the selected microorganism prior to fermentation. When performing acid hydrolysis at a high concentration, substances might appear after neutralization that inhibits subsequent fermentation [25] so this should be taken into account. Another possible approach would be to perform saccharification and fermentation simultaneously in a single step. This process must be combined with dilute acid or high temperature pretreatment; it also requires compatible conditions (pH, temperature, substrate concentration) for the enzymatic treatment and the fermentation process [27].

The ethanol yields obtained cover a very wide range, depending on the microalga strain, the culture conditions, and the hydrolysis and fermentation conditions (**Table 1**); hence all of the variables involved should be optimized for each specific case [20, 24].

The last step of the process is ethanol recovery and ethanol purification from the fermentation broth. This is usually performed using distillation or rectification, which is yet a well-established step for first and second-generation ethanol production processes.

Microalga	Hydrolysis	Fermentation	Ethanol production, $g \cdot g^{-1}$ biomass	Reference
Chlorococcum humicola	Acid (3% v/v H2SO4, 160°C, 15 min)	*Saccharomyces cerevisiae*	0.52	[23]
Chlorella vulgaris	Acid (3% v/v H2SO4, 110°C, 105 min)	*Escherichia coli*	0.40	[28]
Chlamydomonas reinhardtii	Acid (3% v/v H2SO4, 110°C, 30 min)	*Saccharomyces cerevisiae*	0.29	[29]
Scenedesmus obliquus	Acid (2% v/v H2SO4, 121°C, 20 min)	*Zymomonas mobilis*	0.21	[30]
Chlorococcum infusionum	Basic (0.75% w/v NaOH, 120°C, 30 min)	*Saccharomyces cerevisiae*	0.26	[18]
Chlorella vulgaris	Enzymatic (pectinase)	*Saccharomyces cerevisiae*	0.89	[31]
Chlamydomonas reinhardtii	Enzymatic (amylase and glucoamylase)	*Saccharomyces cerevisiae*	0.23	[32]
Chlamydomonas fasciata	Enzymatic (glutase)	*Saccharomyces cerevisiae*	0.19	[33]
Synechococcus sp.	Enzymatic (lysozyme and α-glucanases)	*Saccharomyces cerevisiae*	0.27	[34]

Table 1.
Bioethanol production under different conditions.

3.2 Biodiesel production

In the EU, diesel constitutes around 70% of total transport fuel; among biofuels, biodiesel is the most widely used, accounting for an annual production of 14.3 million m^3 [2]. In the USA, biodiesel production accounts for 72 million m^3 with a total capacity of 95 [35]; this includes fatty acid methyl esters (FAMEs) and diesel blends (HVOs). Biodiesel can be used for transport as it is, or it can be blended with fossil diesel fuel at various dosages up to 7%, according to European legislation [17]. Biodiesel is produced from feedstocks such as soybean oil (the main feedstock in the USA), rape seed oil (the main feedstock in the EU), corn oil, canola oil, used cooking oils, sunflower seed oil, palm oil or animal fats. As with bioethanol, it can be a first, second or third-generation biofuel depending on the source.

Focusing on microalgae as the feedstock, biomass productivities above 30 $g \cdot m^{-2} \cdot d^{-1}$ are easily attainable; however, this is not compatible with high lipid contents (%d wt.). The accumulation of fatty acids can be triggered by modulating the culture conditions, for instance, by inducing N starvation, but this leads to a drop in biomass productivity. As an example, [12] achieved biomass and fatty acid productivities of 33.1 and 3.5 $g \cdot m^{-2} \cdot d^{-1}$, respectively, under stress-free conditions. Even if a modest productivity of 8 $g \cdot m^{-2} \cdot d^{-1}$ is supposed, $29 \cdot 10^3$ $kg \cdot Ha^{-1} \cdot y^{-1}$ of biomass would be produced, which is much higher than the reported values for corn grains, at around 3.2–9.6·10^3 $kg \cdot Ha^{-1} \cdot y^{-1}$ [8]. Therefore, the lipid fraction of the biomass can be destined for biodiesel production under the above-mentioned biorefinery concept.

With regard to the process for biodiesel production from microalgae, it usually starts with cell rupture as a pretreatment to increase accessibility to intracellular

Microalga	Water content, %	Method	Solvents	Lipids extracted, %d wt. biomass	Reference
Nannochloropsis gaditana	—	Solvent extraction (30 ml·g^{-1} biomass, 10 min)	Water:chloroform: methanol (1:1:2)	24.5	[39]
Phaeodactylum tricornutum	—	Solvent extraction (30 ml·g^{-1} biomass, 10 min)	Water:chloroform: methanol (1:1:2)	13.1	[39]
Chaetoceros calcitrans	—	Solvent extraction (30 ml·g^{-1} biomass, 10 min)	Water:chloroform: methanol (1:1:2)	8.7	[39]
Chlorella vulgaris	—	Solvent extraction (12 ml·g^{-1} biomass, 2 h)	Chloroform:methanol (1:2)	28.5	[40]
Nannochloropsis gaditana	—	Solvent extraction (10 ml·g^{-1} biomass, 60°C, 45 min)	Methanol	38.3	[41]
Microalga	—	**Method**	**Solvents**	**FAME yield, %**	**Reference**
Schizochytrium limacinum	80	Solvent extraction (8 ml·g^{-1} biomass, 90 °C, 40 min)	Methanol (catalyst H2SO4)	8.45	[42]
Chlorella pyrenoidosa	90	Solvent extraction (10 ml·g^{-1} biomass, 150°C, 2 h)	Hexane:methanol 6:4 (catalyst H2SO4)	89.8	[43]
Nannochloropsis salina	76	Solvent extraction (7.5 ml·g^{-1} biomass, 100°C, 1 h)	Methanol (catalyst H2SO4)	99.8	[44]

Table 2.
Biodiesel production under different conditions.

lipids. This is followed by lipid extraction using physical and/or chemical methods [36]. Normally, chemical extraction is performed with a single organic solvent, such as hexane, or preferably a mixture of polar and non-polar solvents to increase selectivity and extractability. Lipid extraction by supercritical CO2 is also an interesting alternative that offers multiple advantages [37]; however, at the moment it is not economically feasible at the large-scale. Other novel methods are being studied at the lab-scale, such as ultrasound-assisted extraction, microwave-assisted extraction, and the use of ionic liquids or bio-based solvents [38]. The FAME extraction yield is highly dependent on the biomass composition, the extraction method used, the solvent/s chosen and their ratios, variables such as temperature and process time, and biomass to solvent/s ratios (**Table 2**). Therefore, all these issues need to be optimized in each case.

The extracted lipids can be used as a fuel in power-generating devices but not in vehicle engines because of their low thermo-physical properties and high viscosity. Consequently, once the lipids have been extracted, they can be hydrotreated to produce HVOs or transesterified to obtain FAMEs.

Another alternative is to perform direct biomass transesterification to obtain biodiesel in a single step, thus enabling the use of wet biomass, which would avoid the high costs attached to biomass drying. The drawback is that this process requires very intensive treatment so it hinders the valorization of other biomass fractions under a biorefinery concept. The microalgal biomass is processed with a suitable solvent and catalyst (usually H2SO4) at a high temperature (the higher the temperature, the lower the reaction time required). If wet biomass is used, one must take into account that the presence of water will have a negative effect on the catalyst's activity and its interaction with the lipids. For example, [43] found that it was necessary to increase the temperature from 90 to 150°C in order to maintain a high FAME conversion yield when the biomass contained 90% moisture. Furthermore, the use of wet biomass requires the addition of a greater amount of extraction solvent.

3.3 Biogas production

Biogas is produced from the anaerobic digestion of organic matter. The technology for doing this is well established. Besides sewage, manure, agricultural residues and other wastes, microalgal biomass can be used as the feedstock, therefore obtaining a third-generation biofuel. Anaerobic digestion is a complex sequence of metabolic reactions (hydrolysis, acidification, acetogenesis and methanogenesis) that allows the biomass to be converted into biogas by means of an anaerobic bacteria consortium. Biogas is composed of methane (50–75%), carbon dioxide (around 25–500%) and small amounts of other gases such as nitrogen and oxygen [45]. After upgrading to biomethane, it can be used in transport as Liquid Natural Gas (LNG) or Compressed Natural Gas (CNG). Biogas can also be combusted to generate heat/electricity. Biogas production worldwide was $59{\cdot}10^3$ million m^3 in 2014. The EU leads this production with 29 10^3 million m^3, followed by China with 15 and then the USA with 9 [2].

When using microalgal biomass as the feedstock, a pretreatment is often required, mainly when the cell walls are composed of cellulose or hemicellulose, which are very hard and/or very thick. Once again, physical (ultrasound, high-pressure homogenization, heating...), chemical (oxidation, basic treatment...) or enzymatic treatments can be selected for this purpose [46]. Biodegradability mainly depends on the biochemical composition and the cell wall so anaerobic digestion is highly species-dependent. Moreover, it is preferable to use wet biomass as dry biomass diminishes biogas production by about 20% [47]. It is also important to pay attention to the biomass' C:N ratio since a low value reduces its digestibility—the optimal value is around 20–35. A high protein content makes the process difficult as it triggers the generation of ammonia, which has toxic effects on the bacteria. This is what happens when the biomass is previously subjected to lipid extraction to obtain biodiesel and the residual biomass is used for anaerobic digestion. To increase the C:N ratio, co-digestion with another raw material can be performed.

However, other inhibitory factors usually influence the process such as the presence of sulfate/sulfide, volatile fatty acids or high salinity.

Anaerobic digestion of the biomass is usually carried out at temperatures of 25–38 or 50–55°C [48], no higher as this would also inhibit biogas production. Ehimen et al. [49] found a 61% greater CH4 yield at 35°C when compared to 25°C. Ward et al. [50] found faster biogas production when working at 55°C, achieving a 95% methane yield in only 11 days. However, this process requires higher energy inputs. With regard to the hydraulic retention time in the system, this should be maintained at around 30 days as it has been shown that methane production tends to rise asymptotically to a maximum value reached at around 30 days [51].

Taking all the above into account, we can see that the biomass conversion yield into biogas varies considerably (**Table 3**). A maximum biogas yield of 611 $L{\cdot}kg^{-1}$ volatile solids (ash-free biomass) was reported for a biomass composed of several microalgae species [55].

After anaerobic digestion, a separation step is carried out. A solid residue is obtained that can be used as a biofertilizer and a liquid stream that can be recirculated into the cultivation system. This provides the nitrogen, phosphorous and other nutrients required to cultivate the biomass, which helps to make the process sustainable. Accordingly, [56] recycled 40.7 g of nitrogen (74%) and 3.8 g of phosphorous (35%) from 1 kg d wt. of biomass.

3.4 Biocrude production

A different approach is to perform a thermochemical conversion of the whole microalgal biomass into hydrocarbon fuels by thermochemical conversion. If the biomass is dry, the processes for this would be torrefaction, pyrolysis and gasification for solid, liquid and gas feedstocks, respectively. These processes could technically be an option but, as mentioned before, it is mandatory to work with wet biomass to make the process economically viable. Thermochemical processes for wet biomass are hydrothermal carbonization, hydrothermal liquefaction and hydrothermal gasification for solid, liquid and gas feedstocks, respectively. Of these, hydrothermal liquefaction (HTL) seems to be the most promising [57]. The chemistry of the process is still not known in full but it involves three main steps: depolymerization, decomposition and recombination. The process consists

Microalga	Reaction time, d	T, °C	Biogas production, $m^3{\cdot}kg^{-1}$ volatile solids	Methane content, %	Reference
Arthrospira maxima	30	35	0.200	72	[52]
Chlamydomonas reinhardtii	32	38	0.587	66	[47]
Dunaliella salina	32	38	0.505	64	[47]
Chroococcus sp.	30	36	0.487	55	[53]
Euglena gracilis	32	38	0.485	67	[47]
Arthrospira platensis	32	38	0.481	61	[47]
Chlorella vulgaris	32	30	0.467	75	[54]
Chlorella kessleri	32	38	0.335	65	[47]
Scenedesmus obliquus	32	38	0.287	62	[47]

Table 3.
Biogas production for several microalgae with no pre-treatment.

Microalga	Reaction time, min	T, °C	Biocrude yield, %	Reference
Tetraselmis sp.	5	350	65	[61]
Nannochloropsis sp.	60	350	43	[62]
Chlorella pyrenoidosa	30	280	64	[63]
Chlamydomonas reinhardtii	60	230	71	[64]
Desmodesmus sp.	5	375	49	[65]

Table 4.
Biocrude production for several microalgae.

of heating wet biomass up to 370°C and 25 MPa for less than 30 min [58]. The bio-oil obtained is later separated from the aqueous phase. Subsequently, it has to be upgraded to liquid fuel and then refined to jet fuel, diesel or gasoline in order to make it suitable for use in transportation. The aqueous phase could be partially recycled to the algae photobioreactor [59] or to the hydrothermal reactor [60]. Alternatively, it could be upgraded to fuel gas (CO_2 and CH_4) by catalytic hydrothermal gasification.

The conversion yield of microalgal biomass into biocrude by HTL depends on the biomass loading, temperature, pressure, reaction time and catalysts used; values close to 70% have been reported (**Table 4**). This product provides a heating value of up to 39 MJ·kg^{-1} [66], similar to that from petroleum crude but containing large amounts of nitrogen and sulfur that lead to higher NO_x and SO_x emissions and greater viscosity and oxidation processes. Selecting adequate homogeneous/heterogeneous catalysts to increase the conversion yield and improve the quality of the obtained biocrude is being researched in depth.

4. Biorefinery concept

Microalgae production facilities must operate under a biorefinery concept in order to obtain as many bioproducts as possible in addition to the biofuels. As explained before, the production of biofuels from microalgae is technically but not economically feasible. It is essential to lower the biomass production cost by improving the photobioreactors design and the biomass harvesting. It is also important to optimize extraction processes based on wet biomass, as drying imposes one of the greatest production costs. Furthermore, the process should be coupled with wastewater treatment and the use of CO2 from exhaust gases to reduce costs; however, this would limit the end use of the obtained products because of contamination.

With regard to the process chain, one should bear in mind that this depends, to a large extent, on the species and culture conditions employed. Therefore, it is not possible to propose a general process flow diagram as this depends on the strain, its biochemical composition, and if it produces a specific compound of interest as a bioproduct, etc. In whichever case, the first step is to break down the cells (mechanical, chemical or enzymatic processes). Then, the main objective is to separate each fraction of interest without damaging the others. For example, some authors propose using the protein fraction of the biomass first, followed by the carbohydrates and finally the lipids [67]. This would preserve the quality of the amino acids, which would otherwise be diminished by the steps involved in carbohydrate and lipid valorization. It was then proposed to use the carbohydrates first, followed by the proteins and finally the lipids as the process was more efficient [68]. Further approaches are available in the literature [69], some recommending the valorization of the carbohydrates first [70] while others start with lipid valorization [71]. In any case, the residual biomass will finally undergo an anaerobic digestion process to obtain biogas and recycle nutrients to the system.

5. Conclusions

Nowadays, microalgae production is considered as part of a biorefinery concept. The challenge is to obtain as many bioproducts from the biomass as possible—the production of biodiesel (from the lipid fraction), bioethanol (from the carbohydrate fraction) and biogas (from the residual biomass) or biocrude (from the whole biomass)—all part of a biomass valorization chain. These products will not replace current fuels but, together, they will help to improve transport sustainability when used in conjunction with other renewable energies.

Acknowledgements

This work has been funded by the AL4BIO Project from the Spanish Ministry of Science, Innovation and Universities and the SABANA Project from the European Union's Horizon 2020 Research and Innovation Program.

Conflict of interest

The authors declare no conflict of interest.

Author details

Cynthia V. González-López*, Francisco García-Cuadra, Natalia Jawiarczyk, José M. Fernández-Sevilla and Francisco G. Acién-Fernández
Department of Chemical Engineering, University of Almería, Almería, Spain

*Address all correspondence to: cynthiagonzalez@ual.es

References

[1] Ellabban O, Abu-Rub H, Blaabjerg F. Renewable energy resources: Current status, future prospects and their enabling technology. Renewable and Sustainable Energy Reviews. 2014;**39**:748-764

[2] WBA. WBA Global Bioenergy Statistics 2019; 2017

[3] Dutta K, Daverey A, Lin J-G. Evolution retrospective for alternative fuels: First to fourth generation. Renewable Energy. 2014;**69**:114-122

[4] Mata TM, Martins AA, Caetano NS. Microalgae for biodiesel production and other applications: A review. Renewable and Sustainable Energy Reviews. 2010;**14**(1):217-232

[5] Hu Q. Environmental effects on cell composition. In: Richmond A, editor. Handbook of Microalgal Culture: Biotechnology and Applied Phycology. Oxford: Blackwell Publishing Ltd; 2004. pp. 83-93

[6] Dasan YK, Lam MK, Yusup S, Lim JW, Lee KT. Life cycle evaluation of microalgae biofuels production: Effect of cultivation system on energy, carbon emission and cost balance analysis. Science of the Total Environment. 2019;**688**:112-128

[7] Pittman JK, Dean AP, Osundeko O. The potential of sustainable algal biofuel production using wastewater resources. Bioresource Technology. 2011;**102**(1):17-25

[8] IEA, "State of Technology Review-Algae Bioenergy," 2017.

[9] European Commission. DIRECTIVE (EU) 2015/1513 of the European Parliament and of The Council of 9 September 2015 amending Directive 98/70/EC relating to the quality of petrol and diesel fuels and amending Directive 2009/28/EC on the promotion of the use of energy from renewables; 2015

[10] San Pedro A, González-López CV, Acién FG, Molina-Grima E. Outdoor pilot-scale production of Nannochloropsis gaditana: Influence of culture parameters and lipid production rates in tubular photobioreactors. Bioresource Technology. 2014;**169**:667-676

[11] San Pedro A, González-López CV, Acién FG, Molina-Grima E. Outdoor pilot production of nannochloropsis gaditana: Influence of culture parameters and lipid production rates in raceway ponds. Algal Research. 2015;**8**:205-213

[12] San Pedro A, González-López CV, Acién FG, Molina-Grima E. Outdoor pilot production of Nannochloropsis gaditana: Influence of culture parameters and lipid production rates in flat-panel photobioreactors. Algal Research. 2016;**18**:156-165

[13] Jorquera O, Kiperstok A, Sales EA, Embiruçu M, Ghirardi ML. Comparative energy life-cycle analyses of microalgal biomass production in open ponds and photobioreactors. Bioresource Technology. 2010;**101**(4):1406-1413

[14] Tredici MR et al. Energy balance of algal biomass production in a 1-ha 'Green Wall panel' plant: How to produce algal biomass in a closed reactor achieving a high net energy ratio. Applied Energy. 2015;**154**:1103-1111

[15] Mussatto SI et al. Technological trends, global market, and challenges of bio-ethanol production. Biotechnology Advances. 2010;**28**(6):817-830

[16] RFA. 2019 Ethanol Industry Outlook. Powered with Renewed Energy. Washington D.C.; 2019

[17] Baudry G, Macharis C, Vallée T. Can microalgae biodiesel contribute to achieve the sustainability objectives in the transport sector in France by 2030? A comparison between first, second and third generation biofuels though a range-based multi-actor multi-criteria analysis. Energy. 2018;**155**:1032-1046

[18] Harun R, Danquah MK, Forde GM. Microalgal biomass as a fermentation feedstock for bioethanol production. Journal of Chemical Technology and Biotechnology. 2010;**85**:199-203

[19] Gouveia L. Microalgae as a Feedstock for Biofuels. Berlin: Springer-Verlag Berlin Heidelberg; 2011

[20] Doan QC, Moheimani NR, Mastrangelo AJ, Lewis DM. Microalgal biomass for bioethanol fermentation: Implications for hypersaline systems with an industrial focus. Biomass and Bioenergy. 2012;**46**:79-88

[21] Fernández JFS, González-López CV, Fernández FGA, Sevilla JMF, Grima EM. Utilization of anabaena sp. in CO_2 removal processes: Modelling of biomass, exopolysaccharides productivities and CO_2 fixation rate. Applied Microbiology and Biotechnology. 2012;**94**(3):613-624

[22] Singh A, Olsen SI, Nigam PS. A viable technology to generate third-generation biofuel. Journal of Chemical Technology and Biotechnology. Nov. 2011;**86**(11):1349-1353

[23] Harun R, Danquah MK. Influence of acid pre-treatment on microalgal biomass for bioethanol production. Process Biochemistry. 2011;**46**(1):304-309

[24] Daroch M, Geng S, Wang G. Recent advances in liquid biofuel production from algal feedstocks. Applied Energy. 2013;**102**:1371-1381

[25] de Farias Silva CE, Bertucco A. Bioethanol from microalgae and cyanobacteria: A review and technological outlook. Process Biochemistry. 2016;**51**:1833-1842

[26] Harun R, Yip JWS, Thiruvenkadam S, Ghani WAWAK, Cherrington T, Danquah MK. Algal biomass conversion to bioethanol—A step-by-step assessment. Biotechnology Journal. 2014;**9**(1):73-86

[27] Phwan CK, Ong HC, Chen WH, Ling TC, Ng EP, Show PL. Overview: Comparison of pretreatment technologies and fermentation processes of bioethanol from microalgae. Energy Conversion and Management. 2018;**173**:81-94

[28] Lee S, Oh Y, Kim D, Kwon D, Lee C, Lee J. Converting carbohydrates extracted from marine algae into ethanol using various ethanolic Escherichia coli strains. Applied Biochemistry and Biotechnology. 2011;**164**:878-888

[29] Nguyen MT, Choi SP, Lee J, Lee JH, Sim SJ. Hydrothermal acid pretreatment of Chlamydomonas reinhardtii biomass for ethanol production. Journal of Microbiology and Biotechnology. 2009;**19**(2):161-166

[30] Ho S-H, Li P-J, Liu C-C, Chang J-S. Bioprocess development on microalgae-based CO2 fixation and bioethanol production using Scenedesmus obliquus CNW-N. Bioresource Technology. 2013;**145**:142-149

[31] Kim KH, Choi IS, Kim HM, Wi SG, Bae H-J. Bioethanol production from the nutrient stress-induced microalga Chlorella vulgaris by enzymatic hydrolysis and immobilized yeast fermentation. Bioresource Technology. 2014;**153**:47-54

[32] Choi SP, Nguyen MT, Sim SJ. Enzymatic pretreatment of Chlamydomonas reinhardtii biomass for ethanol production. Bioresource Technology. 2010;**101**(14):5330-5336

[33] Asada C, Doi K, Sasaki C, Nakamura Y. Efficient extraction of starch from microalgae using ultrasonic homogenizer and its conversion into ethanol by simultaneous Saccharification and fermentation. Natural Resources. 2012;**3**:175-179

[34] Möllers KB, Cannella D, Jørgensen H, Frigaard N-U. Cyanobacterial biomass as carbohydrate and nutrient feedstock for bioethanol production by yeast fermentation. Biotechnology for Biofuels. 2014;7(1):64

[35] EIA. Monthly Biodiesel Production Report. With Data for April 2019. Washington DC; 2019

[36] Mubarak M, Shaija A, Suchithra TV. A review on the extraction of lipid from microalgae for biodiesel production. Algal Research. 2015;**7**:117-123

[37] Taher H, Al-Zuhair S, Al-Marzouqi AH, Haik Y, Farid M. Enzymatic biodiesel production of microalgae lipids under supercritical carbon dioxide: Process optimization and integration. Biochemical Engineering Journal. 2014;**90**:103-113

[38] Deshmukh S, Kumar R, Bala K. Microalgae biodiesel: A review on oil extraction, fatty acid composition, properties and effect on engine performance and emissions. Fuel Processing Technology. 2019;**191**:232-247

[39] Ríos SD, Castañeda J, Torras C, Farriol X, Salvadó J. Lipid extraction methods from microalgal biomass harvested by two different paths: Screening studies toward biodiesel production. Bioresource Technology. 2013;**133**:378-388

[40] Kim Y-H et al. Ultrasound-assisted extraction of lipids from Chlorella vulgaris using [Bmim][MeSO4]. Biomass and Bioenergy. 2013;**56**:99-103

[41] Bermúdez Menéndez JM et al. Optimization of microalgae oil extraction under ultrasound and microwave irradiation. Journal of Chemical Technology and Biotechnology. 2014;**89**(11):1779-1784

[42] Johnson MB, Wen Z. Production of biodiesel fuel from the microalga Schizochytrium limacinum by direct Transesterification of algal biomass. Energy & Fuels. 2009;**23**(10):5179-5183

[43] Cao H, Zhang Z, Wu X, Miao X. Direct biodiesel production from wet microalgae biomass of Chlorella pyrenoidosa through In situ Transesterification. BioMed Research International. 2013:6

[44] Kim T-H et al. Development of direct conversion method for microalgal biodiesel production using wet biomass of Nannochloropsis Salina. Bioresource Technology. 2015;**191**:438-444

[45] Kwietniewska E, Tys J. Process characteristics, inhibition factors and methane yields of anaerobic digestion process, with particular focus on microalgal biomass fermentation. Renewable and Sustainable Energy Reviews. 2014;**34**:491-500

[46] Khalid A, Arshad M, Anjum M, Mahmood T, Dawson L. The anaerobic digestion of solid organic waste. Waste Management. 2011;**31**(8):1737-1744

[47] Mussgnug JH, Klassen V, Schlüter A, Kruse O. Microalgae as substrates for fermentative biogas production in a combined biorefinery concept. Journal of Biotechnology. 2010;**150**(1):51-56

[48] Jankowska E, Sahu AK, Oleskowicz-Popiel P. Biogas from

microalgae: Review on microalgae's cultivation, harvesting and pretreatment for anaerobic digestion. Renewable and Sustainable Energy Reviews. 2017;**75**:692-709

[49] Ehimen EA, Sun ZF, Carrington CG, Birch EJ, Eaton-Rye JJ. Anaerobic digestion of microalgae residues resulting from the biodiesel production process. Applied Energy. 2011;**88**(10):3454-3463

[50] Ward AJ, Hobbs PJ, Holliman PJ, Jones DL. Optimisation of the anaerobic digestion of agricultural resources. Bioresource Technology. 2008;**99**(17):7928-7940

[51] Ras M, Lardon L, Bruno S, Bernet N, Steyer J-P. Experimental study on a coupled process of production and anaerobic digestion of Chlorella vulgaris. Bioresource Technology. 2011;**102**(1):200-206

[52] Samson R, Leduy A. Biogas production from anaerobic digestion of Spirulina maxima algal biomass. Biotechnology and Bioengineering. 1982;**24**(8):1919-1924

[53] Prajapati SK, Kaushik P, Malik A, Vijay VK. Phycoremediation and biogas potential of native algal isolates from soil and wastewater. Bioresource Technology. 2013;**135**:232-238

[54] Sánchez Hernández EP, Travieso Córdoba L. Anaerobic digestion of Chlorella vulgaris for energy production. Resources, Conservation and Recycling. 1993;**9**(1-2):127-132

[55] Golueke O. Anaerobic digestion of algae. Applied Microbiology. 1957;**5**:47-55

[56] Zhang Y, Kendall A, Yuan J. A comparison of on-site nutrient and energy recycling technologies in algal oil production. Resources, Conservation and Recycling. 2014;**88**:13-20

[57] López Barreiro D, Ronsse F, Brilman W. Hydrothermal liquefaction (HTL) of microalgae for biofuel production: State of the art review and future prospects. Biomass and Bioenergy. 2013;**53**:113-127

[58] Toor SS, Rosendahl L, Rudolf A. Hydrothermal liquefaction of biomass: A review of subcritical water technologies. Energy. 2011;**36**(5):2328-2342

[59] Biller P et al. Nutrient recycling of aqueous phase for microalgae cultivation from the hydrothermal liquefaction process. Algal Research. 2012;**1**(1):70-76

[60] Ramos-Tercero EA, Bertucco A, Brilman DWF. Process water recycle in hydrothermal liquefaction of microalgae to enhance bio-oil yield. Energy & Fuels. 2015;**29**(4):2422-2430

[61] Eboibi BE, Lewis DM, Ashman PJ, Chinnasamy S. Effect of operating conditions on yield and quality of biocrude during hydrothermal liquefaction of halophytic microalga Tetraselmis sp. Bioresource Technology. 2014;**170**:20-29

[62] Brown TM, Duan P, Savage PE. Hydrothermal liquefaction and gasification of Nannochloropsis sp. Energy & Fuels. 2010;**24**(6):3639-3646

[63] Yang L, Ma R, Ma Z, Li Y. Catalytic conversion of Chlorella pyrenoidosa to biofuels in supercritical alcohols over zeolites. Bioresource Technology. 2016;**209**:313-317

[64] Hognon C et al. Comparison of pyrolysis and hydrothermal liquefaction of Chlamydomonas reinhardtii. Growth studies on the recovered hydrothermal aqueous phase. Biomass and Bioenergy. 2015;**73**:23-31

[65] Torri C, Garcia Alba L, Samorì C, Fabbri D, Brilman DWF. Hydrothermal

treatment (HTT) of microalgae: Detailed molecular characterization of HTT oil in view of HTT mechanism elucidation. Energy & Fuels. 2012;**26**(1):658-671

[66] Mathimani T, Mallick N. A review on the hydrothermal processing of microalgal biomass to bio-oil - knowledge gaps and recent advances. Journal of Cleaner Production. 2019;**217**:69-84

[67] García-Cuadra F, Jawiarczyk N, González-LópezCV,Fernández-SevillaJM, Acien G. Valorización de biomasa de microalgas: Aprovechamiento de proteínas, carbohidratos y lípidos. Revista Latinoamericana de Biotecnología Ambiental y Algal. 2012;**3**(2):147-161

[68] Ortiz Montoya E, Llamas Moya B, Molina Grima E, García Cuadra F, Fernández Sevilla JM, and Acién Fernández FG. Method for the valorisation of photosynthetic microorganisms for integral use of biomass. WO2014/122331; 2014

[69] Zhu L. Biorefinery as a promising approach to promote microalgae industry: An innovative framework. Renewable and Sustainable Energy Reviews. 2015;**41**:1376-1384

[70] Suali E, Sarbatly R. Conversion of microalgae to biofuel. Renewable and Sustainable Energy Reviews. 2012;**16**(6):4316-4342

[71] Morris HJ, Almarales A, Carrillo O, Bermúdez RC. Utilisation of Chlorellavulgaris cell biomass for the production of enzymatic protein hydrolysates. Bioresource Technology. 2008;**99**(16):7723-7729

Chapter 5

Recycling of Waste Plastics into Pyrolytic Fuels and Their Use in IC Engines

Sinan Erdogan

Abstract

The energy crisis and environmental destruction are the principal problems in the present day due to the rapid industrialization and growing population. Degradation of solid waste such as plastic bottles, grocery bags, etc. in nature takes many years. Besides, plastic disposing methods like landfill, reusing, and burning can create severe risks to the human health and environment. Therefore, plastic must be kept under control from damaging the environment. One of the most favorable and effective disposing methods is pyrolysis, which is an environmentally friendly and efficient way. Pyrolysis is the thermal degradation of solid wastes at high temperatures to produce pyrolytic oil. The pyrolytic oil produced is converted into pyrolytic fuel very similar to diesel or gasoline by upgrading. The calorific value of the pyrolytic fuel is similar to that of diesel and gasoline. Pyrolytic fuel can be used in internal combustion engines without significant loss in engine performance. Besides, some engine emissions, especially smoke opacity and CO and HC emissions, improve when used with additives or when the engine's operating conditions such as compression ratio and ignition timing are changed. However, NO_x emission is very similar to diesel fuel, too.

Keywords: recycling of waste, waste plastic oil, pyrolytic fuel, alternative fuel, emissions

1. Introduction

Sustainable mobility is described as "the ability to meet society's need to move freely, gain access, communicate, trade and establish relationships without sacrificing other essential human or ecological values, today or in the future" by the World Business Council for Sustainable Development [1]. The main purpose of sustainable mobility approach is to promote mobility solutions that remarkably decrease greenhouse gas emissions, increase energy efficiency, and decrease traffic congestion, creating safer, healthier, and more livable cities and environments [2]. The sustainable mobility approach focuses on the realization of three basic actions as Avoid-Shift-Improve (ASI) framework, which are as follows: (1) Avoid: making less travel (reduce the need for travel), (2) Shift: changing the mode of transport, and (3) Improve: reducing the length of the journey and using more efficient systems in transportation technologies [2, 3]. The third action within the ASI framework focuses on improved efficiency engines with new-generation fuel technology and alternative fuels that reduce the environmental impact of transport. NDC Global

Outlook Report 2019 of the United Nations (UN) states that greenhouse gas emissions should be reduced by 45% over the next decade compared to 2010 and should be net-zero by 2050 [4]. In the transportation sector, it is desirable that the vehicles in transportation consist of electric vehicles and that the electricity consumed is produced from renewable sources in order to reach these targets. However, replacing present internal combustion engine vehicles with zero-emission electric vehicles and generating electricity from renewable sources is so difficult and costly that it will not be possible in a short- and medium-term. As the short- and medium-term activity, it is aimed to reduce greenhouse gas (GHG) emissions below a certain limit by increasing the use of alternative fuels such as biodiesel, bioethanol, and pyrolytic fuels in transportation. These fuels are very important in terms of producing from waste as well as vegetable sources. In addition, the short- and medium-term targets include the development of internal combustion engines with lower fuel consumption and more environmentally friendly.

This chapter will focus on the production of pyrolytic fuel from waste plastics and their use in vehicles. The degradation of plastic materials takes several hundred years and thus, they affect the environment. In landfill areas, waste plastics can release carcinogens and other toxic chemicals that pollute groundwater. In addition, these poisonous chemicals can disrupt soil fertility. Plastic particles floating on the ocean surface are a threat to the marine ecosystem [5]. If waste plastics are used directly as a source of energy and are burned, it creates very harmful emissions to the environment and increases the amount of GHG. In this context, recycling of waste plastic as a fuel through pyrolysis in an inert atmosphere is an environmentally friendly solution. In the last 30years, there has been a rapid growth in the plastics industry. Worldwide production of synthetic polymers such as polyvinyl chloride (PVC), polypropylene (PP), polyethylene (PE), and polystyrene (PS) has increased more than 100 times in the last 3 years. These plastics are widely used in daily important applications such as household appliances, clothing, food packaging, electronics, and automotive products. While plastics bring great convenience to our lives, the processing of waste plastics is inevitable and must be solved soon [6].

The recycling of plastic wastes is a considerable matter nowadays. Many researchers have investigated the applicability of using waste plastic oil in a diesel engine due to the decrease in fossil resources in the world and the increasing amount of plastic waste. It has been concluded that waste plastic oil has similar properties to diesel fuel and can be used instead of diesel [7, 8]. In this chapter, pyrolysis, one of the alternative fuel production methods, is examined. Types of pyrolysis, types of waste plastics as raw materials, parameters affecting pyrolysis, and products obtained from pyrolysis are presented. In some experimental studies in the literature, the physical and chemical properties of pyrolytic fuels are compared with commercial diesel and gasoline fuels. In the last part, experimental studies on the use of waste plastic pyrolysis oil as fuel in internal combustion engines are given. Waste plastic pyrolysis oil as fuel was evaluated in terms of engine performance, emission, and combustion characteristics.

2. What is pyrolysis?

Pyrolysis is the thermal decomposition of biomass at high temperatures in an inert atmosphere. The process cannot be reversed due to the chemical composition of the material changes. The word pyrolysis is derived from the Greek word pyro "fire" and lysis "separating" [9]. The products formed at the end of pyrolysis include solid coal (char), liquid pyrolysis oil, and gas. Pyrolytic liquid and gas products can be used for power generation in engines and turbines. Therefore, the pyrolysis of biomass is a promising method for alternative energy sources. This method takes place at high temperatures. The critical temperature ranges for obtaining pyrolysis products is between 350 and 500°C. The process can use self-produced pyrolytic oil or gas to reach such high temperatures [10]. In this respect, the process can be performed by consuming less energy from the outside. During the pyrolysis process, the chemical bonds of the biomass are thermally degraded in an oxygen-free environment and at high temperatures.

Advantages of pyrolysis over other methods are as follows: the amount of harmful wastes released to the environment is low, the end product formation is economical, and it reduces air pollution and supports itself in terms of energy use. Primary products obtained in the pyrolysis process can be used directly or can be converted to high-quality fuel or other chemical products by applying chemical processes. This method is often preferred for the conversion of solid biomass waste, which is difficult to use and cost-effective, into liquid products. The production, transportation, and storage costs of these liquid products are low and their energy density is high. Besides, in the elemental composition of liquid products, there are complex structures of biomass-like and oxidized hydrocarbons.

2.1 Type of pyrolysis

Pyrolysis is divided into three basic groups: slow, flash (intermediate), and fast according to the time and temperature during the process. According to the pyrolysis types given in **Table 1**, the liquid products obtained, especially from the fast pyrolysis processes can be used as internal combustion engine fuel [11].

Slow pyrolysis is slowly heating of organic biomass in the absence of oxygen at nearly 500°C. Instead of combusting, the volatiles from the organic material vaporize partially, and a product called charcoal remains, consisting of a large proportion (approximately 80%) of carbon. Slow pyrolysis is also named carbonization. In the slow pyrolysis, the main product is solid charcoal [12].

Fast pyrolysis is rapidly heating of organic biomass often in the range between 425 and 500°C in the absence of air. Organic vapors, char, and gas are produced under these conditions. In later stages of the process, the vapors are condensed to pyrolytic oil. Commonly, 60–75 wt.% of the feedstock is converted into oil [12].

Flash pyrolysis is a thermal-cracking process at a very high heating rate with a very short vapor residence time with high pyrolysis temperatures around 450 and 1000°C. The main objective of this process is to minimize secondary cracking. Thus, the liquid yields are maximized and can go up to 75% [13].

2.2 Types of plastics as the raw material

In general, different types of plastics have different compositions. Plastic products are manufactured differently to meet different needs. The different types of plastics and their uses are shown in **Table 2**.

Pyrolysis types	Retention time	Rate of heating	Temperature (°C)	Liquid yield (%)
Slow	5–30 min	Low	300–650	<30
Fast	<2 s	Very high	450–600	50–75
Flash	<1 s	High	450–1000	>75

Table 1.
The operating parameters and product yields for pyrolysis types [13–16].

Code	Type of plastics	Common packaging applications
1 PETE	Polyethylene terephthalate (PET)	Soft drink, water, and vegetable oil bottles; peanut, butter, and jam jars; salad containers
2 HDPE	High-density polyethylene (HDPE)	Milk bottles, yogurt and margarine tubs, waste bags, bowls, cable insulations, liquid detergents, motor oils, shampoos, and perfume containers
3 V	Polyvinyl chloride (PVC)	Juice bottles, cling films, soft toys, electrical insulations, roofing materials, pipes and window frame materials
4 LDPE	Low-density polyethylene (LDPE)	Frozen food bags; squeezable bottle, e.g., honey, mustard; cling films; flexible container lids
5 PP	Polypropylene (PP)	Margarine and yogurt tubes, ketchup bottles, bags for chips and biscuits, disposable cups and plates, medicine bottles, chairs
6 PS	Polystyrene (PS)	Egg cartoons; disposable cups, plates, trays, and cutlery; disposable take-away containers; yogurt and margarine containers
7 OTHER	Other	Beverage bottles; baby milk bottles

Table 2.
The different types of plastics and their uses.

The chemical composition of the plastic compound includes moisture, fixed carbon, volatile matter, and ash. **Table 3** presents the proximate analysis data of different plastics. The change in biomass composition affects the yields of pyrolysis products. If the amount of volatile matter is high in the plastic content, the liquid yield increases, while the ash content is high, the char yield increases. The ash content is considered low for all plastics, while the volatile matter is very high. These properties mean that plastics may have a high potential to generate a high yield of liquid oil during the pyrolysis process according to the operating conditions [17].

According to ultimate analysis of feedstocks, the carbon (C) ratio by weight in the content of PET is 63.94%. The C ratio of HDPE, PVC, LDPE, PP, and PS is 86.99, 37.24, 85.6, 86.88, and 91.57%, respectively [18]. High carbon content indicates that it will turn into end products with high calorific value.

For plastics, depending on their chemical structure, thermal decomposition begins to occur at different temperatures. For common plastics such as PET, HDPE, LDPE, PP, and PS, the thermal decomposition temperature starts at 350°C, while the thermal decomposition temperature of PVC starts at a temperature lower than 220°C. It is also necessary to adjust the operating temperatures according to the final product preference. For example, if it is desired to produce mostly gas and char, temperatures higher than 500°C are recommended; if it is preferred to produce mostly liquid, it is recommended to apply at temperatures between 300 and 500°C. PET and PVC are rarely investigated by researchers because they produce very low oil yield compared to other types of plastics. Pyrolysis is not recommended for some types of plastics because they contain harmful substances and have low yields such as PVC [19].

Plastic types of polyethylene terephthalate (PET) have become an excellent selection for plastic packaging for a variety of food products, especially drinks such as fruit juice, soft drink bottles, and mineral water cans. In some experimental studies, PET has been used as a raw material in the pyrolysis process with the fixed-bed reactor at 500°C at a heating rate of 10 and 6°C/min. The liquid product yields are 23.1 wt and 39.89 wt%, gas yields are 76.9 wt and 52.13 wt%, and char yields are 0 and 8.98 wt% in these studies [28, 29].

High-density polyethylene (HDPE) is a very strong and economical material. It is not used in products where transparency is important because it has a milk color appearance. Due to its low cost, ease of forming, and break resistance, it has a wide usage area. HDPE is widely used in the producing of detergent bottles,

Type of plastics	Proximate analysis (wt%)			
	Moisture	Fixed carbon	Volatile	Ash
Polyethylene terephthalate (PET)	<0.7	6–14	85–92	<0.1
High-density polyethylene (HDPE)	<0.3	~0	94–99.8	<1.5
Polyvinyl chloride (PVC)	<0.8	5–7	85–94.8	<0.1
Low-density polyethylene (LDPE)	<0.3	~0	99–99.8	<0.4
Polypropylene (PP)	<0.4	<1.2	95–99.6	1–4
Polystyrene (PS)	<0.3	<0.2	99–99.8	<0.5
Other	<0.2	<3	97–99.8	<1

Table 3.
Proximate analysis of plastics [18, 20–27].

milk bottles, toys, oil containers, and more because of its high strength property. In an experimental study, HDPE has been used as a raw material in the pyrolysis process at 350°C using a semibatch reactor. The liquid product yield is 80.88 wt%. In other study at 550°C, the liquid product yield is 79.08 wt% using a semibatch reactor [30, 31].

Polyvinyl chloride (PVC) is available as transparent and semicrystalline material, depending on heat treatment. The most important usage advantage is that it is fully recyclable. It is hard and impact resistant. Thanks to the chlorine (Cl) in the PVC content, it becomes an excellent fire-resistant material and is therefore very suitable for electrical insulation. However, PVC produces hydrochloric acid (HCl) harmful during the pyrolysis [17]. In an experimental study in which pyrolysis of PVC was performed, it is stated that the liquid yield was 12.79 wt%, and hydrogen chloride gas (HCl) yield was found to be 58.2 wt% among all gases (the yield of all gases was 87.7%) [32].

Plastic types of low-density polyethylene (LDPE) have perfect water resistance, so they are widely applied for plastic bags, garbage bags, wrapping foils for packaging, and much more. In some experimental studies in which maximum liquid yield is obtained, LDPE has been used as a raw material in the pyrolysis processes, which are in a fixed-bed reactor at 500°C with a heating rate of 10°C/min, and in a batch reactor at 550°C with a heating rate of 5°C/min. According to these studies, the liquid yields are found at 95 and 93.1%, respectively [33, 34]. In another study, the liquid yield of LDPE was found to be 74.7 wt% using batch reactor at 430°C with a heating rate of 3°C/min [35].

Polypropylene (PP) is a material resistant to chemicals, heat, and extreme fatigue. It is one of the plastics with medium hardness and gloss. PP has a lower density than HDPE but has higher hardness and rigidity that makes it preferable in the plastic industry. In some experimental studies carried out from 300 to 500°C, the liquid yields increased from 69.82 to 82.12% [29, 31, 36, 37].

Polystyrene (PS) is a versatile and multipurpose plastic, very hard, brittle, glossy, and foam-shaped plastic. It is an inexpensive resin with a relatively low melting point. PS offers reasonable durability, strength, and lightness. For this reason, it is used in a variety of sectors such as construction, electronics, medical appliances, food packaging, and toy. In some experimental studies carried out from 425 to 600°C, the liquid yields obtained were between 89.5 and 98.7% [38–40].

2.3 By-products of plastic pyrolysis

The primary products obtained from pyrolysis can be used without treatment. In addition, they can be converted to secondary products after product improvement techniques to improve fuel properties and can be used as a more efficient fuel. In particular, the liquid product obtained from pyrolysis can be converted into a fuel of very similar properties to diesel or gasoline fuel after secondary processes. Thus, the internal combustion engine becomes available as fuel.

The gaseous product of plastic pyrolysis contains mostly carbon dioxide, carbon monoxide, hydrogen, ethane, methane, ethylene, etc. The pyrolysis product gases may provide heat for the pyrolysis reactor, which is part of the system, or may be used for the generation of heat and electricity in a gas turbine combined cycle system.

The solid product that has a high surface area and large pores and that is released as a result of the pyrolysis process applied to biomass can be used as activated carbon. In other words, the content of char is composed of condensed organic residues and the inorganic phases, with an average high heating value of 28.5–29 MJ/kg [41].

2.4 Operating parameters of pyrolysis

In pyrolysis, the feedstock is heated to a certain temperature at a given heating rate without oxygen and held there for a certain time. During pyrolysis, large hydrocarbon molecules are broken down into relatively smaller ones by reactions such as depolymerization, dehydration, decarbonylation, decarboxylation, deoxygenation, oligomerization, and aromatization [41]. The quantity and proportion of pyrolysis products are influenced by many factors such as biomass composition, pyrolysis temperature, heating rate, and catalyst effect. The fact findings in the 1980s have shown that the yield of pyrolysis oil increases when a biomass feedstock is heated quickly and the produced vapors condense rapidly [12]. These factors and their effects on the pyrolysis process are described below.

2.4.1 Heating rate

The yields of the products obtained in the pyrolysis process vary depending on the heating rate. It is very important to adjust the heating rate to the optimum level as it affects heat transfer. Increasing the heating rate raises the liquid yields and decreases the char yields. The higher the heating rate, the faster the volatiles form, and the lower the heating rate, the longer the residence time for the volatiles. Thus, repolymerization reactions that form char occurs. Higher heating rates minimize char formation. At lower heating rates, the main product is char, which works for a long time (several days). This process is called carbonization.

As the heating rate increases, the conversion degree of the final products and the yield of gases increase, while the yields of oil and char decrease [41]. **Figure 1** shows the typical change in the yields of the final products when there is variation of the heating rate during the pyrolysis process.

Most of the studies in the literature have similar trends like in **Figure 1**; however, the product yields vary according to the raw material used. For example, in a study, the pyrolysis of waste plastic bags in a fixed bed reactor is examined at a pyrolysis temperature of 450°C and the heating rate from 5 to 15°C/min. At heating rates of 5, 10, and 15°C/min., liquid yields are 53.1, 51.3, and 47%, while the yields

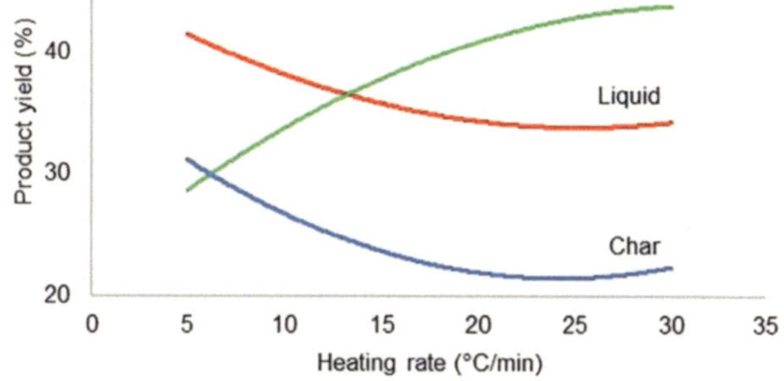

Figure 1.
Yield of the pyrolysis product at various heating rates.

of gases are 32.4, 37.3, and 41%, respectively. Besides, the yields of char are 14.5, 11.4, and 12%, respectively [42].

2.4.2 Temperature

The temperature at which pyrolysis occurs is the most important factor in product distribution. Liquid yield depends on the reaction temperature, biomass type, hot vapor residence time, and catalyst use. It is important to note that the maximum efficiency is not the same as the maximum quality. If the quality of the liquid needs to be optimized, the operating parameters of the process must be carefully defined [16]. According to **Figure 2**, as the temperature increases, the char yield decreases while the yield of the gases increases.

In addition, as the temperature increases up to 500°C, the liquid yield increases. The higher the temperature, the more liquid products turn into gases. After the maximum liquid yield is obtained in the temperature range of 550–550°C, the liquid yield decreases with increasing temperature.

Mostly liquid contains water, oil, and wax. Oil in liquid can be converted into fuel via chemical processes by considering the physical and chemical properties of the fuel. In this case, the oil yield may be slightly lower than the initial liquid yield. In the literature, the changes in temperature of pyrolysis products show similar tendencies. In a study by Pütün [43], bio-oil yields of 41, 46, and 43% were obtained at temperatures 400, 550, and 700°C, respectively.

The yield of the product obtained from the pyrolysis of the plastic mixture in another study, according to the variation of temperature, is shown in **Table 4** [44].

It is clear that the gas percentage increases as the temperature increases. The yield of the char decreases with increasing temperature. It is stated that it supports more gas formation as molecules break down under high temperature and heating rate conditions. In addition, they form much smaller organic molecules under these conditions. When a higher amount of energy is available at a higher temperature, there is a tendency for an increasing number of secondary reactions. The amount of oil and wax decreases with an increase in temperature. At low temperatures, the conversion reactions of solid products to pyrolytic oil decrease and waxy structures increase due to incomplete pyrolysis. Cahyono and Fenti [42] conducted an experimental study using plastic wastes in a reactor at the heating rate of 15°C/min with pyrolysis temperature between 250 and 450°C. **Table 5** shows the yield of the pyrolysis products obtained in the study. Accordingly, the amount of wax decreases with increasing temperature, while the amount of oil and gases increases.

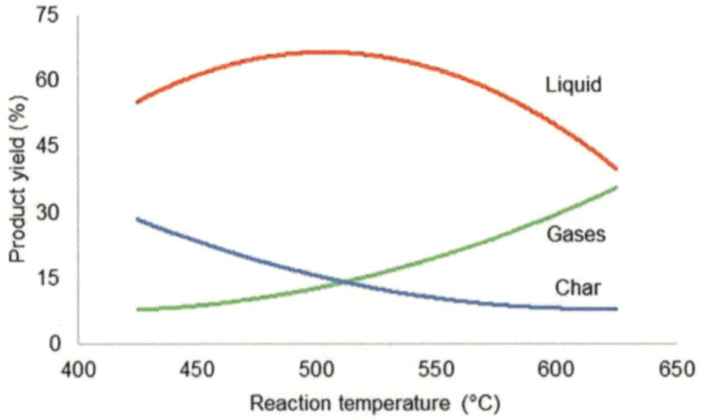

Figure 2.
Yield of the pyrolysis product at various reaction temperatures.

Product (%)	Temperature (°C)				
	500	550	600	650	700
Gas	9.79	24.52	43.33	88.76	68.81
Wax	17.28	18.56	8.72	0	0
Oil	37.79	38.55	34.44	20.49	18.44
Char	2.82	5.87	7.59	–*	–*

Unavailable.

Table 4.
The product yields from the pyrolysis of the plastic mixture with change in temperature.

Product (%)	Temperature (°C)		
	250	350	450
Gas	5	35	41
Wax	52	27	2
Oil	8	15	45
Char	35	23	12

Table 5.
The product yields from the pyrolysis of the plastic with change in low temperature [42].

When the temperature reaches a higher degree, the initially formed liquid products are transformed into gas via secondary reactions. The main part of the product is solid at low temperatures, while the main product is gas at high temperatures [44].

2.4.3 Particle size

In the literature, it is stated that particle size is significantly effective in the generation of pyrolysis products, especially coal and oil. In general, as the particle size increases, the core temperature of the particle is lower than that of the surface. This results in higher temperature gradients within the particle. Therefore, it takes a long time to complete the pyrolysis process. If fast or flash pyrolysis is to be carried out, it is necessary to divide the biomass into small particles. As a result, char yield increases with increasing particle size and oil and gas yield increases with decreasing particle size. In addition, small biomass particle sizes are needed to achieve high biomass heating rates [16]. Most of the studies in the literature have similar trends like in **Figure 3** for particle size in the pyrolysis process. The yield of the products may vary depending on the raw material.

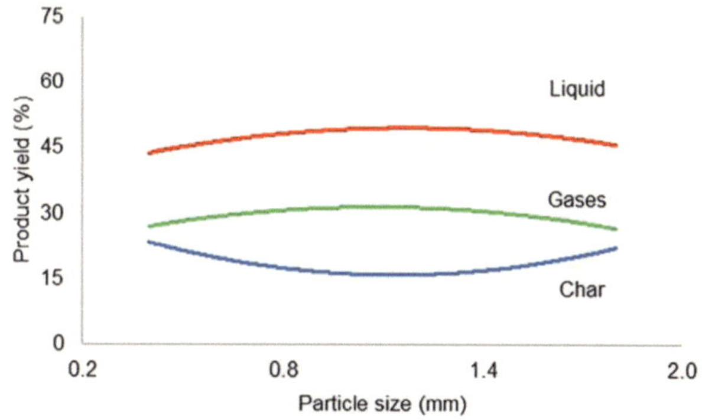

Figure 3.
Yield of the pyrolysis product at various particle sizes.

In the literature, the experimental studies show that particle size does not cause a significant change in char, oil, and gas yield. Ertas and Alma [45] compared the yields of pyrolysis products according to different particle sizes such as 1.60 > Ps > 0.850, 0.850 > Ps > 0.420, and 0.420 > Ps > 0.250 mm in their experimental study (Ps: particle size). It was carried out in a fixed-bed reactor at the optimum temperature (500°C). Nonetheless, the optimum particle size for maximum bio-oil yield is shown as the medium particle size. The gas yield for the large particle size was measured slightly higher than the other dimensions, and the char yield for the small particle size was measured slightly lower than the other dimensions.

2.4.4 Residence time

Volatile residence time is a significant factor to affect yields of liquid and gaseous products in a biomass. A longer residence time allows the formation of secondary reactions such as thermal cracking, repolymerization, and recondensation, thereby reducing liquid yield [46]. Higher temperatures and longer volatile residence times enhance biomass conversion to gas. Mild temperatures and short volatile residence time are optimum for liquids [12].

2.4.5 Catalyst effect

The catalyst is used to accelerate the chemical reaction. In the process using the catalyst, the activation energy of the process is reduced and thus the reaction rate is increased. This case reduces the optimum temperature required for the process. The most important parameter that prevents the application of the pyrolysis process is high energy consumption. So, the use of a catalyst can help save energy [17]. The reaction time is shortened by using a catalyst in the pyrolysis of plastic biomass. Panda carried out pyrolysis of three different types of plastics (PP, LDPE, and HDPE) by changing catalyst to biomass ratios of 1:0, 1:20, 1:10, and 1:3 at 500°C reaction temperature. The reaction time has been shorter when the ratio of catalyst to biomass has increased in his study. When HDPE has been used as biomass, the reaction time without catalyst has been 83 minutes, while the reaction time has been reduced to 64 minutes with 33.3% (1:3) catalyst. The specific gravity of the pyrolysis oil measured 0.853 g/cm^3 when using %5 (1:20) catalyst in the pyrolysis process while it measured 0.789 g/cm^3 when using %5 (1:20) catalyst [47]. By using different catalysts and determining the optimum catalyst/biomass ratio, more chemically homogeneous and different ratios of pyrolysis products can be obtained. In recent years, various biomass-derived waste materials have been investigated using various catalysts to understand the effects of the catalyst in the pyrolysis process.

2.5 Catalysts

The catalysts are widely used in industries and research to optimize product distribution. In the reactor, the use of catalysts for pyrolysis of polyolefins increases the selectivity of volatile products by changing the product distribution [17]. Especially catalysts are of great interest in the production of automotive fuels such as gasoline and diesel [48]. The catalyst has been used by many researchers for product upgrades to improve the hydrocarbon distribution to obtain the pyrolysis liquid, which has similar properties to conventional fuel such as gasoline and diesel.

There are two structures of catalysts, which include only one phase (homogeneous) and more than one phase (heterogeneous). The most common type of catalyst used is heterogeneous because the solid catalyst is easily separated from the liquid product mixture. Heterogeneous catalysts are preferred because the catalysts can be reused [17]. Thus, the use of highly costly catalysts becomes more economical. Zeolite, silica (SiO_2), calcium oxide (CaO), and alumina (Al_2O_3) are the most commonly used catalysts in the pyrolysis process because of reducing the temperature required to complete the reaction and enhance the reaction rate [49].

3. Characteristics of pyrolytic fuel

Plastic waste is now one of the significant elements of municipal solid waste (MSW). It is a mixture of different plastic products, largely made from high-density polyethylene (HDPE), low-density polyethylene (LDPE), polystyrene (PS), polypropylene (PP), polyethylene terephthalate (PET), and polyvinyl chloride (PVC). PS and PE are the most common types of plastics in municipal waste [50]. Plastics are nonbiodegradable polymers mostly containing carbon, hydrogen and few other elements [6]. The elemental analysis of pyrolytic oil and other fuels is given in **Table 6**.

After the oils produced in pyrolysis were removed from visible precipitates, the appearance of the liquid fuel obtained from HDPE is light yellowish. The liquid fuel obtained from PS is seen as dark yellow, while the liquid fuel obtained from PP is seen as deep brown [53].

Density, kinematic viscosity, thermal value, flash point, cetane number, octane number, and cold filter plugging point are among the most important physical and chemical properties of fuels used in internal combustion engines. These characteristics affect the usage decision of fuel in internal combustion engines. Some of the physical and chemical properties of pyrolytic fuels produced from different types of plastics are presented in **Table 7**.

Density is an important fuel feature that affects fuel consumption and sprays characteristics. The densities of liquid fuels produced from HDPE, PS, and PP have been given 0.796, 0.894, and 0.786 g/cm^3, respectively [53].

Kinematic viscosity adjusts the spray pattern and atomization of injected fuel in a combustion chamber [54]. Generally, the fuels with high viscosity are not preferred for internal combustion engines because they lead to poor engine performance. The high viscosity makes it difficult to transport the fuel to the fuel supply system This condition limits the use of high viscosity fuels in winter. In addition, the use of low viscosity fuels can cause serious pump and injector leakage. This may lead to a drop in fuel distribution and a reduction in the engine power output. Low kinematic viscosity demonstrates that pyrolytic oil contains high amounts of gasoline and low amounts of heavy oil. The kinematic viscosities of pyrolytic liquid fuels produced from HDPE, PS, and PP were given as 2373, 1461, and 2115 mm^2/s, respectively [53].

Raw material of fuel	**Elemental analysis (wt%)**					**High calorific value (MJ/kg)**
	Carbon	**Hydrogen**	**Nitrogen**	**Sulfur**	**Others**	
HDPE	85.49	14.23	—	0.28	—	43.92
LDPE	85.44	14.31	—	0.25	—	43.28
PP	83.80	13.85	—	0.33	2.02	38.10
PS	91.48	7.41	—	0.19	0.92	38.53
PVC	39.17	4.96	—	0.58	55.29	22.45
MPW*	84.30	12.25	<0.1	0.38	3.00	39.72
Diesel [51]	85.60	14.10	0.30	—	—	45.19

**Municipal plastic wastes.*

Table 6.
Characteristics of pyrolytic fuels from different plastic types [51, 52].

Properties	PET	HDPE	PVC	LDPE	PP	PS	Gasoline	Diesel
Calorific value (MJ/kg)	28.2	40.5	21.1	39.5	40.8	43.0	42.5	43
Viscosity (mm^2/s)	n.a	5.08[a]	6.36[b]	5.56[c]	4.09[a]	1.4[d]	1.17	1.9–4.1
Density @ 15°C (g/cm^3)	0.90	0.89	0.84	0.78	0.86	0.85	0.78	0.81
Pour point (°C)	n.a	−5	n.a	n.a	−9	−67	—	6
Flash point (°C)	n.a	48	40	41	30	26.1	42	51
Octane number MON (min)	n.a	85.3	n.a	n.a	87.6	n.a	81–85	—
Octane number RON (min)	n.a	95.3	n.a	n.a	97.8	90–98	91.95	—
Diesel index	n.a	31.05	n.a	n.a	34.35	n.a	—	40

n.a. not available.
[a]at 40°C.
[b]at 30°C.
[c]at 25°C.
[d]at 50°C.

Table 7.
Fuel characteristics of plastic pyrolytic fuel [19].

Pour point is known as the temperature at which the fluid stops to flow. Generally, the increase in viscosity may cause the fluid to lose its flow feature. If liquid fuel has a lower pour point, it has lesser paraffin content but greater aromatic content [17].

In order to prevent fire hazard during storage, one of the important features of the fuel is the flashpoint. The flashpoint of the liquid is defined as the lowest temperature at which the liquid-vapor mixed with the air ignites when an external flame is applied [55]. The flashpoint of HDPE, PVC, and LDPE pyrolysis oil was very close to commercial gasoline.

It is desirable to have a high cetane index as an indicator of the excellent combustion properties of the liquid product. The high cetane index is associated with the presence of α-olefins and linear paraffin [31]. Distillation is the process by which a liquid mixture is separated into its components by heating and condensing to a certain temperature. In this process, the initial boiling point (IBP, °C) is determined by the temperature at which the first distillation drops. The final boiling point (FBP, °C) is the temperature at which all the oil sample in the distillation chamber evaporates. The distillation temperatures of 10, 50, 90, and 95% volumes in the recovered volume are recorded as well as the determination of IBP and FBP. Using these temperature data, estimating the calculated cetane index (CCI) of the fuel can be calculated by a four-variable equation according to ASTM D4737 [5]. The boiling temperatures of the different fuels and the mass distribution of the distilled fuels are shown in **Figure 4**.

Pyrolytic oil is compared with diesel, kerosene, and gasoline. It is clear from this figure that pyrolytic oil consists of compounds in the gasoline range, as well as compounds in the range of kerosene and diesel. The starting and ending temperatures of boiling and the thermal values of the conventional fuels are given in **Table 8**. Considering that the liquid at boiling point higher than 150°C is kerosene or diesel, the pyrolytic oil contains approximately 35% gasoline according to **Figure 4**. The rest consists of kerosene and diesel fuel.

The calorific values of some plastics, such as HDPE, PP, and LDPE, have more than 40 MJ/kg [17]. These values may be considered sufficient to be used as fuel. In some studies, the calorific values of HDPE and PP have been reported to be more than 45 MJ/kg, which indicates that these fuels are produced very close to gasoline and diesel fuel [31]. In general, PET and PVC have the lowest calorific value below

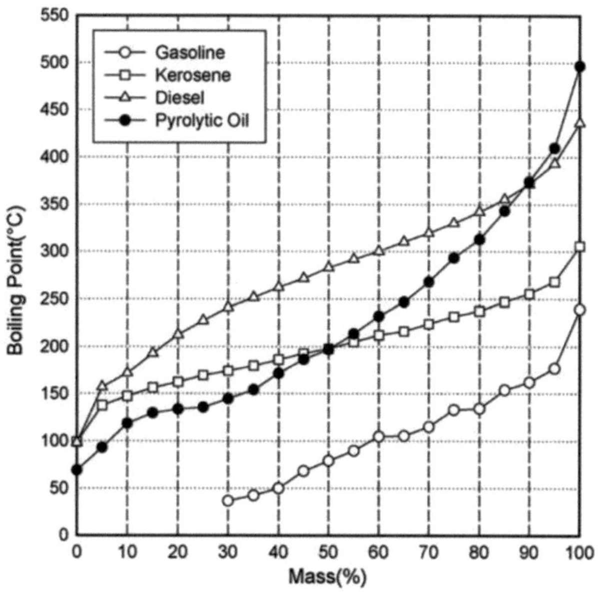

Figure 4.
Distributions of boiling point range of fuel samples [56].

Fuel	Boiling point (°C)	Calorific value (MJ/kg)
Gasoline	40–200	43.4–46.5
Kerosene	150–300	43.0–46.2
Diesel	150–390	42.8–45.8

Table 8.
Boiling point and calorific value of conventional fuels [57].

30 MJ/kg due to the presence of benzoic acid in PVC and chlorine compound in PVC, which reduced fuel quality [17].

The release of hydrochloric acid and chlorine compound during the PVC pyrolysis indicated that the liquid oil was not suitable to be used as a fuel since it depreciated between the fuel quality [17].

The liquid oil that is produced through plastic wastes via pyrolysis may contain certain substances such as sulfur, chlorine, solid residue, moisture, and acids. The presence of these substances not only reduces the quality of the liquid oil but also limits its commercial use. Therefore, liquid oil requires posttreatment, including upgrading to removal of char particles and acids, and neutralization to improve liquid oil with stable pH, and low corrosivity. There are two ways to make the pyrolysis liquid commercially available. One of them is to blend pyrolysis liquid with diesel fuel to provide certain fuel properties. The other is refining. Liquid pyrolysis oil can be used in modified diesel engines after upgrading to fuel and removing from impurities.

4. Use of plastic pyrolysis oil as fuel in the IC engines

Mani et al. [7] have obtained fuel by feeding waste plastics into a reactor along with 1 wt.% catalyst and 10 wt.% coal at a temperature of 300–400°C for about 3–4 hours at atmospheric pressure. In this process, they produced 75% liquid and tested using it in the engine. The engine tests were conducted on a single-cylinder, four-stroke, air-cooled diesel engine at a constant speed of 1500 rpm under 4.4 kW load. The physical and chemical properties of test fuels are as follows: the density, viscosity, high heat value (HHV), and cetane number of the diesel fuel are 0.840 g/cm^3, 2.0 mm^2/s, 46.5 MJ/kg, and 55, respectively, while those of waste plastic oil are 0.836 g/cm^3, 2.52 mm^2/s, 44.34 MJ/kg, and 51, respectively. It is stated that the ignition delay in the test using waste plastic oil was considerably longer than that of the test using diesel. The longer ignition delay in the waste plastic oil caused the cylinder peak pressure for diesel is 67 bar at rated power and 71 bar in the case of waste plastic oil. During the premixed combustion phase, the longer ignition delay resulted in a higher heat release rate usage of waste plastic oil. NO_x emission for waste plastic oil increased by approximately 25% compared to diesel. The amount of NO_x increased when the temperature inside the cylinder rose. Furthermore, the residence time of the fuel in the reactions affected NO_x formation [31]. It is expressed that the high heat release rate increased the exhaust gas temperature and triggered NO_x formation. In the case of waste plastic oil usage, HC emissions increased by approximately 15% compared to diesel. Besides, it is stated that the CO emission of waste plastic oil was higher than that of diesel. The CO_2 emission in the waste plastic oil was generally lower than the CO_2 emission in the diesel, except under full load conditions. The smoke of waste plastic oil decreased by approximately %40 according to that of diesel at rated power. In terms of engine performance, the thermal efficiency was calculated as 28.2% at rated power for diesel and 27.4% for the waste plastic oil.

In another study, Mani et al. [58] collected plastic HDPE grocery bags from local retailers and produced a fuel using a pyrolysis batch reactor. After that, they mixed waste plastic oil and diesel fuel in certain proportions and tested in a diesel engine. The mixture of 10% waste plastic oil and 90% diesel fuel was named WPO10. Similarly, WPO30, WPO50, and WPO70 mixtures were prepared. The WPO30 fuel blend produced results very close to the engine test results obtained from diesel fuel. In this study, it is stated that it can be used by mixing diesel and waste plastic oil.

Gungor et al. [59] produced the waste polyethylene fuels via the pyrolysis process in their study. Thermal cracking experiments were realized for 1 hour over 400°C. Waste polyethylene fuel obtained from experiment is named WPE. The physical and chemical properties of test fuels are as follows: the density, viscosity, calorific value, and cetane number of the diesel fuel are 0.833 g/cm^3, 2.52 mm^2/s, 45.1 MJ/kg, and 54.6, respectively, while those of WPE are 0.788 g/cm^3, 2.33 mm^2/s, 45.5 MJ/kg, and 43.7, respectively. This fuel was mixed with conventional diesel fuel with volumetric ratios of 5, 10, 15, 20, and 100%. In the experiments, a diesel engine with four-stroke, four cylinders at 1800 rpm was used. The engine performance and exhaust emissions of the pyrolytic fuels were determined and they were compared with those of diesel. Power output of the engine in WPE5 blend increased maximum by 1.63% compared with diesel. Torque output of the engine in WPE5 usage decreased by 2.73% compared to diesel, especially at higher engine speeds. While CO emission decreased by 20.63% when WPE5 blend was used and CO_2 emission increased by 3.34% when WPE5 blend was used. Besides that, NO_x emission increased by 9.17% with WPE5 usage compared to diesel.

Guntur et al. [60] investigated the performance and emission characteristics of a diesel engine with single-cylinder, constant speed, and direct injection using waste plastic pyrolysis oil blends as an alternative fuel. The waste plastic pyrolysis oil (WPPO) was blended with diesel fuel with volumetric ratios of 50 and 70%. Results indicated that the brake thermal efficiency was higher compared to diesel at part load condition. HC (except for 100% load), CO, and CO_2 emissions of WPPO50 and WPPO70 fuels were found higher than diesel fuel at all loads.

Kaimal and Vijayabalan [61] collected waste plastics from the municipal disposal site and separated them into small chips (0.5–1 cm2). They were mixed with 10 wt% coal and 1 wt% silica catalyst to perform pyrolysis. At the end of the pyrolysis process, the yield of plastic oil is 80% by input weight. The yield of solid coke residue is 15% by weight and the yield of gaseous fractions (a mixture of propylene, isobutane, ethane, and methane) are 5% by weight. The experiment was conducted on direct injection, single-cylinder, water-cooled diesel engine at 1500 rpm generating 3.7 kW. Tests were conducted for diesel fuel, plastic oil mixtures (PO25, PO50, and PO75), and neat plastic oil (PO100). Due to the higher oxygen content and the higher calorific value of the plastic oil, the maximum heat release is significantly increased for plastic oil and its blends. It was reported that the highest thermal efficiency was 31.46% for diesel at full load. In addition, it is stated that the brake thermal efficiency of PO25, PO50, and PO75 mixtures are 30.07, 29.17, and 28.26%, respectively.

Kalargaris et al. [62] conducted an experimental study by producing via a fast pyrolysis process using a feedstock consisting of different types of plastic. Feedstock composition consists of styrene-butadiene (47%), polyester (26%), clay (12%), ethylene-vinyl acetate (7%), rosin (6%), polyethylene (1%), and polypropylene (1%). Density, kinematic viscosity, flashpoint, and LHV of produced plastic pyrolysis oil (PPO) are 0.9813 g/cm^3, 1.918 mm^2/s, 13°C, and 38.3 MJ/kg, respectively. The properties of diesel control fuel are 0.8398 g/cm^3, 2.62 mm^2/s, 59.5°C, and 42.9 MJ/kg, respectively. PPO was tested in a four-cylinder, turbocharged, direct injection, water-cooled diesel engine. The experiments were carried out at the rated engine speed of 1500 rpm and at different engine loads from 25–100%. Five blending ratios of PPO and diesel, namely 25, 50, 75, 90, and 100% (v/v%), were tested at each load. They reported that cylinder peak pressures enhanced with increasing PPO content in the blend at full load. The lower viscosity of the PPO improves the atomization, thus resulting in increased HRR values of PPO blends. In addition, the oxygen content (3.3 wt.%) in the PPO may contribute to the high HRR of PPO blends. In the study, it was stated that BSFC increased and BTE decreased with an increasing amount of PPO in the blend. In addition, BSFC decreased with increasing engine load and BTE increased. A similar trend was observed in all mixtures. Increased percentage of PPO in the blend also increased EGT. NO_x emission increased both with increasing engine load and increasing PPO ratio in the blend. The CO emission of the PPO blends was measured very close to that of the diesel at high engine loads, while the CO emission was very high in blends with a high PPO ratio at low loads. UHC and CO_2 emissions are higher in PPO blends than in diesel.

Kumar et al. [63] in their study evaluated performance and emission analysis of blends of waste plastic oil obtained by catalytic pyrolysis of waste high-density polyethylene with diesel in a CI engine with varying loads. The physical and chemical properties of test fuels are as follows: the density, viscosity, and gross calorific value of the diesel as control fuel are 0.83 g/cm^3, 2.58 mm^2/s, and 43.8 MJ/kg, respectively, while those of waste plastic oil are 0.79 g/cm^3, 2.1 mm^2/s, and 40.17 MJ/kg, respectively. The experimental results show that the brake thermal efficiencies at all load conditions are lower as compared to those of diesel fuel and exhaust gas temperature increases with increase in engine load. The BSFC increases with increase in WPO blend ratio and decreases with increase in engine load. Mechanical efficiency increases with increasing brake power for all fuel blends. The NO_x emission and CO emission increase with increase in the percentage of waste plastic oil in blends, and NO_x emission decreases while CO emission increases with increase in engine load. The unburned hydrocarbon emission decreases with increase in the engine load and increases with an increase in the percentage of waste plastic oil in blends. The CO_2 emission for the blends is lower than diesel for almost all loads and all blends.

In a study, Mani and Nagarajan [64] showed that the emissions can be reduced by making some modifications in the engine. In the tests carried out with waste plastic oil, the conditions where the standard injection timing of the engine is 23° bTDC and the delayed injection timing is 14° bTDC were compared. After this modification, if the waste plastic oil is used in the engine, it can be possible to obtain results that are equivalent to the results of diesel usage at standard injection timing [7, 64]. When waste plastic oil was used in engine tests with standard injection timing, the ignition delay was longer than that of engine tests with delayed injection timing. A longer ignition delay caused the cylinder peak pressure to rise. It is stated that the cylinder peak pressure is low for the delayed injection timing. The cylinder peak pressure in the engine test with delayed injection time decreased by 6% compared to the engine test with standard injection time at full load. The heat release rate in the delayed injection time decreased by 6% compared to standard injection time at full load. At full load, the brake specific fuel consumption in standard injection timing diminished due to the long ignition delay. The reduction of the brake specific fuel consumption in delayed injection timing led to its high thermal efficiency. The thermal efficiency is 28.2% for standard injection timing, while it is 32.25% for the retarded injection timing. The high thermal efficiency indicates less heat loss. The exhaust gas temperature for the delayed injection time is lower than that of the other. This is explained by the fact that there is less heat loss in the study. The emission of NO_x for retarded injection timing decreased at all loads by approximately 11% to standard injection timing. Because lower peak pressures result in lower peak temperatures, it is seen that retarded injection timing of the engine diminishes the unburned HC emissions. The unburned HC emissions in the engine test with delayed injection time dramatically reduced compared to the engine test with standard injection time. It is explained that the CO emissions were lower for delayed injection timing by 25% compared to standard injection timing in an engine fueled with waste plastic oil. However, the smoke for the delayed injection timing increased compared to the standard injection timing.

Damodharan et al. [5] produced fuel from waste plastics via pyrolysis and tested it at 5.2 kW rated power and 1500 rpm constant speed in a diesel engine with single-cylinder, water-cooled, and direct injection. The effects of injection timing and EGR ratio of the engine were investigated in tests using waste plastic oil (WPO). Properties of test fuels include the following: for diesel, the lower heat value (LHV) is 41.82 MJ/kg, the viscosity is 3.8 mm^2/s, the density is 0.838 g/cm^3, the cetane number is 54, and the flashpoint is 70°C, while for the WPO, they are 40.35 MJ/kg, 2.16 mm^2/s, 0.813 g/cm^3, 51, and 38°C, respectively. It is stated that the standard injection timing of the engine is 23° bTDC and tests are performed at 25°, 23°, and 21° bTDC injection timing. In addition, 10, 20, and 30% EGR rate was used in the tests using neat WPO and compared with the results of diesel at standard injection timing. The peak in-cylinder pressures and heat release rates (HRRs) were reported to be dropping gradually as the injection timing was delayed from 25° bTDC to 21° bTDC. In early injection, waste plastic oil had sufficient time to mix with air and produce a better mixture for combustion. Thus, it resulted in higher pressure and better combustion leading to HRR peaks. In addition, early injection prolongs ignition delay and improves combustion. In this way, fuel consumption drops. It is seen that the minimum fuel consumption is at 25° bTDC. Brake specific fuel consumption (BSFC) steadily enhanced with rising EGR rates at all injection timings. The trend was similar at all injection timings. Brake thermal efficiency (BTE) diminished with increasing EGR rates at all injection timings. For instance, BTE deteriorated from 34.7 to 28.6%, when the EGR rate was increased from 10–30% at 23° bTDC. The trend was similar to both early and late injection timings. BTE of the engine with WPO used at the early injection timing of 25° bTDC under 10% EGR rate was seen to be better than diesel at standard injection timing without EGR by 5.1%. NO_x emissions were lower than diesel usage at standard condition (standard injection timing of 23° bTDC without EGR) when the injection timing delayed at all EGR rates. HC, CO, and smoke decreased with the delay of injection timing. However, these emissions increased with the increase of EGR. HC emission with WPO used at the early injection timing (25° bTDC) under 10% EGR rate was lower than diesel usage at standard condition. However, in other conditions, it was higher than diesel. CO and smoke emissions with WPO used at the early injection timing (25° bTDC) under 10 and 20% EGR rates were lower than diesel usage at standard conditions. As a result, it is expressed that WPO fuel usage at 10% EGR rate in early injection timing increases performance and reduces emissions, instead of diesel usage under standard conditions.

Ramesha et al. [65] carried out pyrolysis using plastic waste as a raw material without oxygen and at a high temperature of about 250–300° C. In addition, they obtained biodiesel from microalgae. They mixed B20 (20% algae oil methyl ester +80% diesel) fuel and waste plastic oil and used it in experiments. The new fuel mixture was named B20AOME10WPO (10% waste plastic oil +90% B20 algae biodiesel). A direct-injection, single-cylinder, four-stroke, and water-cooled diesel engine was used to perform the engine experiments. The motor was run at 1500 rpm and the load was gradually changed between 0 and 100%. Performance, emission, and combustion characteristics were observed and reported. At full load, the BTE was observed to be higher according to diesel control fuel for plastic oil-biodiesel-diesel blend and biodiesel-diesel blend by 15.7 and 12.9%, respectively. At full load, the BTE was observed to be higher according to diesel control fuel for plastic oil-biodiesel-diesel blend and biodiesel-diesel blend by 15.7 and 12.9%, respectively. For plastic-biodiesel-diesel blend, the emission of UBHC and CO decreased when plastic-biodiesel-diesel blend used. And NO_x emission of plastic-biodiesel-diesel blend increased to pure diesel.

Ananthakumar et al. [66] carried out pyrolysis of waste plastics in the absence of atmosphere. The test fuels were prepared with 2.5, 5, and 7.5% by volume of waste plastic oil, 2.5% diethyl ether (DDE), and the remaining by volume of diesel fuel. All blends with DDE additives are named as P2.5, P5, P7.5, and other test fuels are diesel and P100 (100% waste plastic oil). The density, caloric value, flash point, viscosity, and cetane number of the diesel fuel are 0.840 g/cm^3, 44.8 MJ/kg, 50°C, 2 mm^2/s, and 42, respectively. These properties of P100 are 0.790 g/cm^3, 43.34 MJ/kg, 42°C, 2.52 mm^2/s, and 51, respectively. These properties of DEE are also 0.714 g/cm^3, 33.86 MJ/kg, −40°C, 0.23 mm^2/s, and 126, respectively. Experimental study for waste plastic oils and mixtures was made with a variable compression ratio engine. The results of the test for waste plastic oil and its blends were compared with diesel fuel. It has been shown in the study that waste plastic oil can be used as an alternative to diesel at higher compression ratios without any modification to the engine. Mixing waste plastic oil and diesel with additive diethyl ether gave similar results to the diesel. CO_2 emission reduced for WPO and its blends on comparing with diesel. Nevertheless, the CO emission increased. Unburned hydrocarbon and smoke emission were close to diesel in the case of P2.5, P7.5, and P12.5 use. The maximum pressure and rate of heat release for P2.5 and P7.5 were almost similar to those of diesel. Due to the higher rate of heat release and combustion temperature, NO_x emission of WPO and its blends were high. Therefore, waste plastic oil can be blended with diesel and additives like DEE to get comparable results with diesel at higher compression ratios.

5. Conclusion

The rapid increase in plastic production in recent years means that new plastic wastes, which may harm the environment in the coming years, will increase rapidly.

Although plastic products make our lives easier, being insoluble in nature for many years will cause serious environmental problems. The elimination of environmental pollutants with more environmentally friendly solutions is the main issue of sustainable development. In this context, recycling of plastic wastes should be considered in a way that both pollute the environment less and help to consume fewer petroleum resources. At high temperatures and without oxygen, plastics in the municipal solid waste can be recycled using the pyrolysis process and more than 75% of the initial weight can be obtained as pyrolytic oil. The pyrolytic oil can be produced in a very similar composition to diesel or gasoline by some distillation processes. In addition, fuel properties can be improved by some upgrade methods. As a result, a pyrolytic fuel can be produced that can be conveniently used in internal combustion engines. However, when this pyrolytic fuel produced from plastic wastes is used directly in the engine, there is a slight decrease in performance and combustion characteristics. Exhaust emissions are slightly increased compared to standard diesel fuel. In order to overcome these drawbacks, diesel performance additives such as diethyl ether (DDE) may be added to the pyrolytic fuel. Furthermore, engine performance and combustion characteristics can be improved by altering engine operating conditions such as ignition timing and compression ratio. In addition, the exhaust emissions of pyrolytic fuel can be reduced relative to diesel fuel. Consequently, pyrolytic fuels produced from plastic wastes can be used as an engine fuel to pollute the environment less.

Author details

Sinan Erdogan
Adapazari Vocational School, Sakarya University, Sakarya, Turkey

*Address all correspondence to: sinanerdogan@sakarya.edu.tr

References

[1] Williams K. Spatial Planning, Urban Form and Sustainable Transport. Hampshire: Routledge; 2017

[2] Pollock J. Scaling up Solutions for Sustainable Mobility. United Kingdom: Shell Foundation; 2012

[3] Banister D. The sustainable mobility paradigm. Transport Policy. 2008;**15**(2):73-80

[4] Doyle A. NDC Global Outlook Report 2019: The Heat is on Taking Stock of Global Climate Ambition. Bonn: United Nations; 2019

[5] Damodharan D, Sathiyagnanam A, Rana D, Kumar BR, Saravanan S. Combined influence of injection timing and EGR on combustion, performance and emissions of DI diesel engine fueled with neat waste plastic oil. Energy Conversion and Management. 2018;**161**:294-305

[6] Naima K, Liazid A. Waste oils as alternative fuel for diesel engine: A review. Journal of Petroleum Technology and Alternative Fuels. 2013;**4**(3):30-43

[7] Mani M, Subash C, Nagarajan G. Performance, emission and combustion characteristics of a DI diesel engine using waste plastic oil. Applied Thermal Engineering. 2009;**29**(13):2738-2744

[8] Devaraj J, Robinson Y, Ganapathi P. Experimental investigation of performance, emission and combustion characteristics of waste plastic pyrolysis oil blended with diethyl ether used as fuel for diesel engine. Energy. 2015;**85**:304-309

[9] Sayin C, Erdoğan S, Balki MK. Alternatif Yakıtlar: Çevresel Atıkların İçten Yanmalı Motor Yakıtı Olarak Değerlendirilmesi. Ankara: Nobel Akademik Yayıncılık; 2019. p. 35

[10] Demirbas A. Combustion characteristics of different biomass fuels. Progress in Energy and Combustion Science. 2004;**30**(2):219-230

[11] Balat M, Balat M, Kırtay E, Balat H. Main routes for the thermo-conversion of biomass into fuels and chemicals. Part 1: Pyrolysis systems. Energy Conversion and Management. 2009;**50**(12):3147-3157

[12] Mohan D, Pittman CU Jr, Steele PH. Pyrolysis of wood/biomass for bio-oil: A critical review. Energy and Fuels. 2006;**20**(3):848-889

[13] Zaman CZ, Pal K, Yehye WA, Sagadevan S, Shah ST, Adebisi GA, et al. Pyrolysis: A sustainable way to generate energy from waste. In: Samer M, editor. Pyrolysis. Rijeka: IntechOpen; 2017

[14] Jahirul M, Rasul M, Chowdhury A, Ashwath N. Biofuels production through biomass pyrolysis—A technological review. Energies. 2012;**5**(12):4952-5001

[15] Patni N, Shah P, Agarwal S, Singhal P. Alternate strategies for conversion of waste plastic to fuels. ISRN Renewable Energy. 2013;**2013**:902053

[16] Bridgwater T. Challenges and opportunities in fast pyrolysis of biomass: Part I. Johnson Matthey Technology Review. 2018;**62**(1):118-130

[17] Sharuddin SDA, Abnisa F, Daud WMAW, Aroua MK. A review on pyrolysis of plastic wastes. Energy Conversion and Management. 2016;**115**:308-326

[18] Sharuddin SDA, Abnisa F, Daud WMAW, Aroua MK. Energy recovery from pyrolysis of plastic waste: Study on non-recycled plastics (NRP) data as the real measure of plastic waste.

Energy Conversion and Management. 2017;**148**:925-934

[19] Sharuddin S, Abnisa F, Daud W, Aroua M, editors. Pyrolysis of plastic waste for liquid fuel production as prospective energy resource. In: IOP Conference Series: Materials Science and Engineering. IOP Publishing; 2018

[20] Zannikos F, Kalligeros S, Anastopoulos G, Lois E. Converting biomass and waste plastic to solid fuel briquettes. Journal of Renewable Energy. 2013;**2013**:360368

[21] Heikkinen J, Hordijk JD, de Jong W, Spliethoff H. Thermogravimetry as a tool to classify waste components to be used for energy generation. Journal of Analytical and Applied Pyrolysis. 2004;**71**(2):883-900

[22] Ahmad I, Khan MI, Ishaq M, Khan H, Gul K, Ahmad W. Catalytic efficiency of some novel nanostructured heterogeneous solid catalysts in pyrolysis of HDPE. Polymer Degradation and Stability. 2013;**98**(12):2512-2519

[23] Park SS, Seo DK, Lee SH, Yu T-U, Hwang J. Study on pyrolysis characteristics of refuse plastic fuel using lab-scale tube furnace and thermogravimetric analysis reactor. Journal of Analytical and Applied Pyrolysis. 2012;**97**:29-38

[24] Aboulkas A, El Bouadili A. Thermal degradation behaviors of polyethylene and polypropylene. Part I: Pyrolysis kinetics and mechanisms. Energy Conversion and Management. 2010;**51**(7):1363-1369

[25] Jung S-H, Cho M-H, Kang B-S, Kim J-S. Pyrolysis of a fraction of waste polypropylene and polyethylene for the recovery of BTX aromatics using a fluidized bed reactor. Fuel Processing Technology. 2010;**91**(3):277-284

[26] Abnisa F, Daud WW, Sahu J. Pyrolysis of mixtures of palm shell and polystyrene: An optional method to produce a high-grade of pyrolysis oil. Environmental Progress and Sustainable Energy. 2014;**33**(3):1026-1033

[27] Yao D, Yang H, Chen H, Williams PT. Co-precipitation, impregnation and so-gel preparation of Ni catalysts for pyrolysis-catalytic steam reforming of waste plastics. Applied Catalysis B: Environmental. 2018;**239**:565-577

[28] Çepelioğullar Ö, Pütün AE. Utilization of two different types of plastic wastes from daily and industrial life. Journal of Selçuk University Natural and Applied Science. 2013;**2**(2):694-706

[29] FakhrHoseini SM, Dastanian M. Predicting pyrolysis products of PE, PP, and PET using NRTL activity coefficient model. Journal of Chemistry. 2013;**2013**:487676

[30] Kumar S, Singh R. Recovery of hydrocarbon liquid from waste high density polyethylene by thermal pyrolysis. Brazilian Journal of Chemical Engineering. 2011;**28**(4):659-667

[31] Ahmad I, Khan MI, Khan H, Ishaq M, Tariq R, Gul K, et al. Pyrolysis study of polypropylene and polyethylene into premium oil products. International Journal of Green Energy. 2015;**12**(7):663-671

[32] Miranda R, Yang J, Roy C, Vasile C. Vacuum pyrolysis of PVC I. kinetic study. Polymer Degradation and Stability. 1999;**64**(1):127-144

[33] Bagri R, Williams PT. Catalytic pyrolysis of polyethylene. Journal of Analytical and Applied Pyrolysis. 2002;**63**(1):29-41

[34] Marcilla A, Beltrán M, Navarro R. Thermal and catalytic pyrolysis of

polyethylene over HZSM5 and HUSY zeolites in a batch reactor under dynamic conditions. Applied Catalysis B: Environmental. 2009;**86**(1-2):78-86

[35] Uddin MA, Koizumi K, Murata K, Sakata Y. Thermal and catalytic degradation of structurally different types of polyethylene into fuel oil. Polymer Degradation and Stability. 1997;**56**(1):37-44

[36] Sakata Y, Uddin MA, Muto A. Degradation of polyethylene and polypropylene into fuel oil by using solid acid and non-acid catalysts. Journal of Analytical and Applied Pyrolysis. 1999;**51**(1-2):135-155

[37] Lee K-H, Noh N-S, Shin D-H, Seo Y. Comparison of plastic types for catalytic degradation of waste plastics into liquid product with spent FCC catalyst. Polymer Degradation and Stability. 2002;**78**(3):539-544

[38] Onwudili JA, Insura N, Williams PT. Composition of products from the pyrolysis of polyethylene and polystyrene in a closed batch reactor: Effects of temperature and residence time. Journal of Analytical and Applied Pyrolysis. 2009;**86**(2):293-303

[39] Liu Y, Qian J, Wang J. Pyrolysis of polystyrene waste in a fluidized-bed reactor to obtain styrene monomer and gasoline fraction. Fuel Processing Technology. 2000;**63**(1):45-55

[40] Demirbas A. Pyrolysis of municipal plastic wastes for recovery of gasoline-range hydrocarbons. Journal of Analytical and Applied Pyrolysis. 2004;**72**(1):97-102

[41] Kabakcı SB, Hacıbektaşoğlu Ş. Catalytic pyrolysis of biomass. In: Samer M, editor. Pyrolysis. Rijeka: IntechOpen; 2017. p. 168

[42] Cahyono MS, Fenti UI. Influence of heating rate and temperature on the yield and properties of pyrolysis oil obtained from waste plastic bag. Conserve: Journal of Energy and Environmental Studies. 2017;**1**(1):1-8

[43] Pütün E. Catalytic pyrolysis of biomass: Effects of pyrolysis temperature, sweeping gas flow rate and MgO catalyst. Energy. 2010;**35**(7):2761-2766

[44] Williams EA, Williams PT. Analysis of products derived from the fast pyrolysis of plastic waste. Journal of Analytical and Applied Pyrolysis. 1997;**40**:347-363

[45] Ertaş M, Alma MH. Pyrolysis of laurel (*Laurus nobilis* L.) extraction residues in a fixed-bed reactor: Characterization of bio-oil and bio-char. Journal of Analytical and Applied Pyrolysis. 2010;**88**(1):22-29

[46] Onay O, Kockar OM. Slow, fast and flash pyrolysis of rapeseed. Renewable Energy. 2003;**28**(15):2417-2433

[47] Panda AK. Thermo-catalytic degradation of different plastics to drop in liquid fuel using calcium bentonite catalyst. International Journal of Industrial Chemistry. 2018;**9**(2):167-176

[48] Elordi G, Olazar M, Lopez G, Amutio M, Artetxe M, Aguado R, et al. Catalytic pyrolysis of HDPE in continuous mode over zeolite catalysts in a conical spouted bed reactor. Journal of Analytical and Applied Pyrolysis. 2009;**85**(1-2):345-351

[49] Syamsiro M, Saptoadi H, Norsujianto T, Noviasri P, Cheng S, Alimuddin Z, et al. Fuel oil production from municipal plastic wastes in sequential pyrolysis and catalytic reforming reactors. Energy Procedia. 2014;**47**:180-188

[50] Miandad R, Barakat M, Aburiazaiza AS, Rehan M, Ismail I,

Nizami A. Effect of plastic waste types on pyrolysis liquid oil. International Biodeterioration and Biodegradation. 2017;**119**:239-252

[51] Mazak DT, Ávila I, Crnkovic PM, Cordoba AYM, Pagliuso JD. The mutagenic potential caused by the emissions from combustion of crude glycerin and diesel fuel. Brazilian Archives of Biology and Technology. 2015;**58**(2):309-317

[52] Heydariaraghi M, Ghorbanian S, Hallajisani A, Salehpour A. Fuel properties of the oils produced from the pyrolysis of commonly-used polymers: Effect of fractionating column. Journal of Analytical and Applied Pyrolysis. 2016;**121**:307-317

[53] Owusu PA, Banadda N, Zziwa A, Seay J, Kiggundu N. Reverse engineering of plastic waste into useful fuel products. Journal of Analytical and Applied Pyrolysis. 2018;**130**:285-293

[54] Miandad R, Nizami A, Rehan M, Barakat M, Khan M, Mustafa A, et al. Influence of temperature and reaction time on the conversion of polystyrene waste to pyrolysis liquid oil. Waste Management. 2016;**58**:250-259

[55] Liaw H-J, Tang C-L, Lai J-S. A model for predicting the flash point of ternary flammable solutions of liquid. Combustion and Flame. 2004;**138**(4):308-319

[56] Lee K-H. Thermal and catalytic degradation of pyrolytic oil from pyrolysis of municipal plastic wastes. Journal of Analytical and Applied Pyrolysis. 2009;**85**(1-2):372-379

[57] Boundy RG, Diegel SW, Wright LL, Davis SC. Biomass Energy Data Book. Tennessee, U.S: Department of Energy; 2011

[58] Mani M, Nagarajan G, Sampath S. Characterisation and effect of using waste plastic oil and diesel fuel blends in compression ignition engine. Energy. 2011;**36**(1):212-219

[59] Güngör C, Serin H, Özcanlı M, Serin S, Aydın K. Engine performance and emission characteristics of plastic oil produced from waste polyethylene and its blends with diesel fuel. International Journal of Green Energy. 2015;**12**(1):98-105

[60] Guntur R, Kumar D, Reddy VK. Experimental evaluation of a diesel engine with blends of diesel-plastic pyrolysis oil. International Journal of Engineering, Science and Technology. 2011;**3**(6):5033-5040

[61] Kaimal VK, Vijayabalan P. A detailed study of combustion characteristics of a DI diesel engine using waste plastic oil and its blends. Energy Conversion and Management. 2015;**105**:951-956

[62] Kalargaris I, Tian G, Gu S. Combustion, performance and emission analysis of a DI diesel engine using plastic pyrolysis oil. Fuel Processing Technology. 2017;**157**:108-115

[63] Kumar S, Prakash R, Murugan S, Singh R. Performance and emission analysis of blends of waste plastic oil obtained by catalytic pyrolysis of waste HDPE with diesel in a CI engine. Energy Conversion and Management. 2013;**74**:323-331

[64] Mani M, Nagarajan G. Influence of injection timing on performance, emission and combustion characteristics of a DI diesel engine running on waste plastic oil. Energy. 2009;**34**(10):1617-1623

[65] Ramesha D, Kumara GP, Mohammed AV, Mohammad HA, Kasma MA. An experimental study on usage of plastic oil and B20 algae biodiesel blend as substitute fuel to diesel engine. Environmental Science and Pollution Research. 2016;**23**(10):9432-9439

[66] Ananthakumar S, Jayabal S, Thirumal P. Investigation on performance, emission and combustion characteristics of variable compression engine fuelled with diesel, waste plastics oil blends. Journal of the Brazilian Society of Mechanical Sciences and Engineering. 2017;**39**(1):19-28

Section 3

Alternative Fuels

Chapter 6

Gaseous Biofuels to Sustainable Mobility

Carlos Repáraz Martín, Ignacio de Godos Crespo, Marcelo F. Ortega Romero and Bernardo Llamas Moya

Abstract

In an energy transition scenario, setting the target date for the year 2050, during which the master lines are established to achieve a 100% renewable energy generation system (both stationary, thermal and mobility and transportation), all studies indicate that this will be based on the so-called renewable energy mix. In relation to energy sources for transport, in this scenario, everything is foreseen the coexistence of fossil energies (natural gas and propane or autogas) to the detriment in favour of other fuels and energies from renewable sources, such as electricity (batteries) and gases of renewable origin (biomethane, hydrogen, and synthesis gas). That renewable gases have, beyond the significant reductions in pollutant emissions, is the complementarity they have with renewable energy sources such as solar and biogas, as will be seen later in the sections dedicated to the generation and production technologies of each of these fuels, where renewable energy sources play a fundamental role.

Keywords: biogas, biomethane, hydrogen

1. Introduction

In an energy transition scenario, setting the target date for the year 2050, for which the master lines are established to achieve a 100% renewable energy generation system (both stationary, thermal and mobility and transportation), all the revised studies indicate that energy sector will be based on the so-called renewable energy mix. In the meantime, biomethane and hydrogen (from renewable energy) could represent an effective strategy to move towards the targets set by the Renewable Energy Directive [1].

With regard to energy sources for transport, a coexistence of fossil energies is foreseen (natural gas and liquefied petroleum gases (LPG) or autogas), to the detriment in favour of other fuels and energies from renewable sources, such as electricity (batteries) and gases of renewable origin (biomethane, hydrogen and synthesis gas).

This chapter will focus on those energy sources based on gas in its different types and origins: natural gas, renewable natural gas or biomethane, hydrogen and synthesis gas. It has been decided not to include the autogas or LPG, because it is a fuel directly derived from petroleum, whose engine and distribution technology, unlike natural gas, is not compatible with gas fuels that could be renewable.

The main support levers that renewable gases have, beyond the significant reductions in pollutant emissions, is the complementarity they have with renewable energy sources such as solar and biogas, as will be seen later in the sections dedicated to the generation and production technologies of each of these fuels, where renewable energy sources play a fundamental role. In some scenarios, renewable resources for mobility could be considered a 'drop-in' fuel, which means they are interchangeable with a particular petroleum derived fuel. In this sense, the inclusion of these resources contributes to the gradual replacement of the fossil fuels without considerable investment in infrastructure.

2. Generation

2.1 Natural gas

Natural gas is a mixture of different hydrocarbons, usually gaseous, that occurs naturally in the subsurface [2]. It usually appears next to oil, at the top of the same deposits, and the composition, like that of crude oil, varies depending on where it comes from.

2.1.1 Origin and formation

The origin of the natural gas comes from the decomposition of organic matter, which took place between 240 and 70 million years ago, during the time when large reptiles and dinosaurs inhabited the planet (Mesozoic Era). This organic matter came from planktonic organisms that accumulated on the seabed of coastal platforms or in shallow basins of ponds and were buried under successive layers of land by the action of natural phenomena.

Most oil fields usually contain liquid and gaseous hydrocarbons [3]. Normally, gases, being less dense than liquid, tend to occupy the upper part of the porous rock, held by the impermeable rock that acts as a seal. Below is oil and below it, large saltwater deposits.

The discovery of natural gas dates back to ancient times in the Middle East. Thousands of years ago, it was found that there were natural gas leaks that set fire when they ignited, giving rise to the so-called 'burning sources'. In Persia, Greece or India, temples were erected for religious practices around these 'eternal flames' [4].

Natural gas was not known in Europe until it was discovered in Britain in 1659, although it was not commercialized until 1790. In 1821, the inhabitants of Fredonia (United States) observed gas bubbles that traced to the surface in a stream. William Hart, considered the 'father of natural gas', excavated the first American natural gas well.

During the nineteenth century, natural gas was almost exclusively used as a source of light. Its consumption remained very localized due to the lack of transport infrastructure that hindered the transfer of large quantities of natural gas over long distances. In 1890, there was an important change with the invention of leakproof joints in gas pipelines.

It was after the Second World War when the use of natural gas grew rapidly as a result of the development of gas pipeline networks and storage systems [5].

2.1.2 Composition

Although its composition varies depending on the reservoir, its main chemical species is 79–97% methane (in molar or volumetric composition), commonly

	Component	Nomenclature	Natural state
Natural Gas	Methane	CH_4	Gas
	Ethane	C_2H_6	Gas
	Propane	C_3H_8	Liquefiable gas
	Butane	C_4H_{10}	Liquefiable gas
	Pentane	C_5H_{12}	Liquid
	Hexane	C_6H_{14}	Liquid
	Nitrogen	N_2	Gas
	Carbonic gas	$CO - CO_2$	Gas
	Hydrogen sulfide	H_2S	Gas
	Hydrogen	H_2S	Gas
	Water	H_2O	Gas

Table 1.
Natural gas composition. Source: Own elaboration from literature.

exceeding 90–95%. It also contains other gases such as ethane (0.1–11.4%), propane (0.1–3.7%), butane (<0.7%), nitrogen (0.5–6.5%), carbon dioxide (<1.5%), impurities (water vapour, sulphur derivatives) and traces of heavier hydrocarbons, mercaptans, noble gases and other gases (**Table 1**) [6].

During the extraction, some gases that are part of its natural composition are treated and separated for different reasons among others: by their low calorific value (such as nitrogen or carbon dioxide), by their dew point in the gas pipelines (by having a low saturation temperature) or by their resistance of liquefaction of gases (such as carbon dioxide, which solidifies when producing liquefied natural gas or LNG).

Propane, butane and other heavier hydrocarbons are separated in order to have an efficient and safe gas combustion. In the same sense, water (steam) is eliminated in order to avoid clogging gas pipelines because at high pressures methane hydrates could be formed. So, sulphur derivatives are purified to very low concentrations to prevent corrosion, odour formation and sulphur dioxide emissions (causing acid rain) after combustion.

Finally, and for security reasons, traces of mercaptans (including methyl mercaptan, CH_4S), which allow olfactory detection in case of leakage, are added for domestic use.

2.1.3 Natural gas reserves

Natural gas reservoirs are usually found at high depths, either on land ('onshore') or under the sea ('offshore'). Natural gas may be found in reservoirs in two states; 'free' or 'associated'. In the 'free' state, the gas is extracted independently, not together with other compounds, and when it is 'associated', it is mixed with hydrocarbons or other gases from the reservoir.

A natural gas reserve becomes a 'proven reserve' when determining the quantity and quality of the natural gas contained in said deposit, its duration being calculated based on the amount of gas it has and an estimate of the expected consumption. Since carrying out this research and resource calculation process in its entirety implies significant investments, it is common that certain reserves are

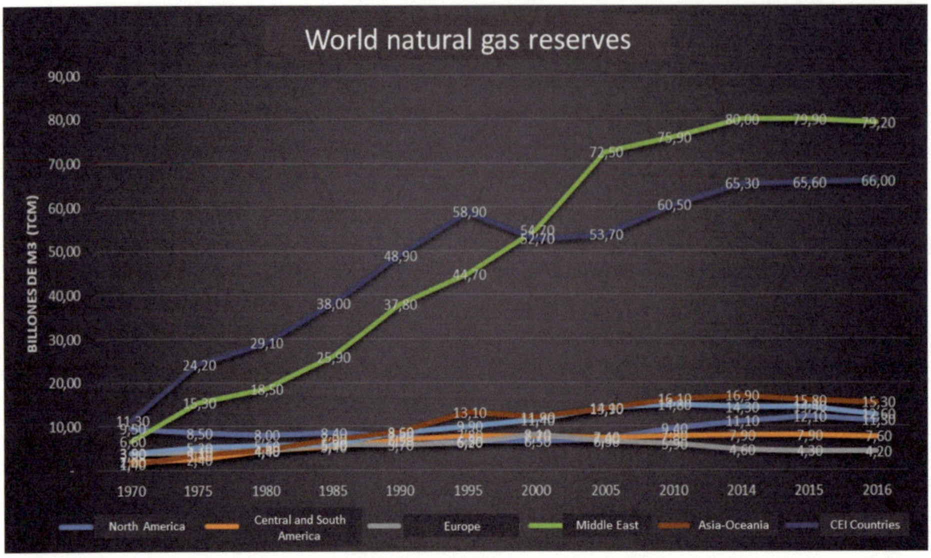

Figure 1.
Proven world reserves of natural gas (source: CEDIGAZ y Oil and Gas Journal. Own elaboration).

only geographically located and their potential is estimated, but they have not been subjected to such precise calculation studies until they are subjected to exploitation. However, gas-producing companies must have demonstrable reserves to guarantee the extraction and supply contracts they incur.

Figure 1 shows the distribution of proven natural gas reserves in the world by geographical area. The highest concentrations of gas are in the Middle East (41% of the world total), followed by the set of CIS countries (Commonwealth of Independent States) representing 31%. European reserves in a downtrend represent 2.7% of total reserves. In the industrial sector, the consumption of natural gas has increased at a faster rate than oil or coal in recent decades; it is estimated that natural gas and electricity will represent about two thirds of the energy used in the industry by 2040 [7].

Following recent research studies and obtaining new extraction technologies and despite the increase in global natural gas consumption, the amount of gas reserves has been increasing. However, new discoveries and the significance of new extraction technologies will have less and less weight, so it is essential to become aware of the efficient use of this resource as well as the research and development of alternative energy sources.

2.2 Biogas/biomethane

Biogas is a gas composed mainly of methane (CH_4) and carbon dioxide (CO_2), in varying proportions depending on the composition of the organic matter from which it was generated. The main sources of biogas are livestock and industrial agro waste, sludge from urban wastewater treatment plants (WWTPs) and the organic fraction of domestic waste. Biogas production through anaerobic digestion can utilise all kinds of organic material except lignin (**Table 2**) [8].

Purified biogas provides reductions in GHG emissions as well as several other environmental benefits when used as a vehicle fuel. Biogas emits less nitrogen oxide, hydrocarbon and carbon monoxide than gasoline or diesel, and engines fuelled by purified biogas are quieter than diesel engines [10].

Component	Livestock wastes	Agricultural wastes	Sewage sludge	Municipal waste	Landfill gas
CH_4	50-80%	50-80%	50-80%	50-70%	45-60%
CO_2	30-50%	30-50%	20-50%	30-50%	40-60%
H_2O	Saturated	Saturated	Saturated	Saturated	Saturated
H_2	0-2%	0-2%	0-5%	0-2%	0-0,02%
H_2S	0-1%	100-700 ppm	0-1%	0-8%	0-1%
NH_3	Traces	Traces	Traces	Traces	0,1-1%
CO	0-1%	0-1%	0-1%	0-1%	0-0,2%
N_2	0-1%	0-1%	0-3%	0-1%	2-5%
O_2	0-1%	0-1%	0-1%	0-1%	0,1-1%

Table 2.
Biogas composition. Source: Ref. [9].

2.2.1 Origin and formation

Biogas is generated from the processes of biological decomposition in the absence of oxygen (anaerobes), which allow biogas to be produced from organic matter, which occur in landfills or in closed reactors commonly known as anaerobic digesters [10]. The degassing of landfills through the capture of the generated biogas allows to improve the safety conditions of exploitation of said landfills, also taking place in many cases an energy use of the biogas collected. In the case of anaerobic digesters, organic matter (substrates) is fed, and certain operating conditions (residence time, temperature, etc.) are maintained. In order to maximize the production of biogas in digesters, it is usual to mix different types of substrates (co-digestion), allowing for the sufficient nutrient concentration required by anaerobic microorganisms. Using the co-digestion strategy, special care must be devoted to chosen mixtures that allow biological processes without inhibitions.

Anaerobic reactors. Anaerobic reactors consist of containers (usually cylindrical in shape), in which the organic matter called substrate is introduced, which will be digested by the action of anaerobic bacteria. Anaerobic digestion facilities also incorporate a gas container, which can be incorporated into the biodigester itself or independently of it as a gasometer.

Other additional elements that are part of the biodigestion facilities are:

- **Premix containers**: function is to homogenize and stabilize input mixtures, prior to entering the biodigester.

- **Sanitation systems**: certain 'inputs', such as animal by-products (not intended for human consumption category), must be sanitized through pasteurization processes, prior to entering biodigesters.

- **Biogas cleaning and upgrading treatments**: the cleaning process involves the passage of biogas through a series of filters, in which certain contaminant components are trapped (sulfur substances, siloxanes) that can damage the mechanical elements used in their subsequent valorization (engines, boilers). The biogas upgrading is the treatment of capture of the CO_2 contained in the biogas, so that the concentration of CH_4 is increased, transforming the raw biogas into biomethane, reusable as renewable natural gas.

- **Energy recovery of biogas**: depending on the final use of the biogas produced, in those cases in which the biogas is not going to be treated for conversion into biomethane, the biogas can be used for energy recovery in cogeneration engines or turbines, to produce electrical and thermal energy, or directly in boiler for production of thermal energy.

- **Digestate treatment**: the input inputs to the biodigester, after passing through the biodigesters, can be processed for later use as a fertilizer. For this, the digestate is subjected to a series of processes (solid-liquid separation, extraction of mineral components, etc.), through which different fertilizer products are obtained such as solid biofertilizer, nitrogen and phosphorus.

Figure 2 shows the different components of an anaerobic biodigestion plant.

Stages of the anaerobic digestion process. During the anaerobic digestion processes, many types of bacteria are produced, which act during the different stages that make up the methanogenesis process, as described below (**Figure 3**).

Disintegration. Initially large particles of biomaterials are disaggregated, and polymers are dissolved as action of the continuous mixture and moderate temperature.

Hydrolysis. Hydrolysis consists in the process of breaking the longer chains of polymeric organic matter, through the action of enzymes secreted by hydrolytic bacteria. The complexity of this process will depend on the complexity of the organic matter entering the digester as well as the conditions in which the process occurs such as temperature, pH, retention time, the biochemical composition of the substrate, the size of particles, NH_4^+ concentration and the concentration of the hydrolysis products.

Fermentation or acidogenic stage. During this stage the fermentation of the soluble organic molecules takes place in compounds that can be used directly by the methanogenic bacteria (acetic, formic, H_2) and smaller organic compounds

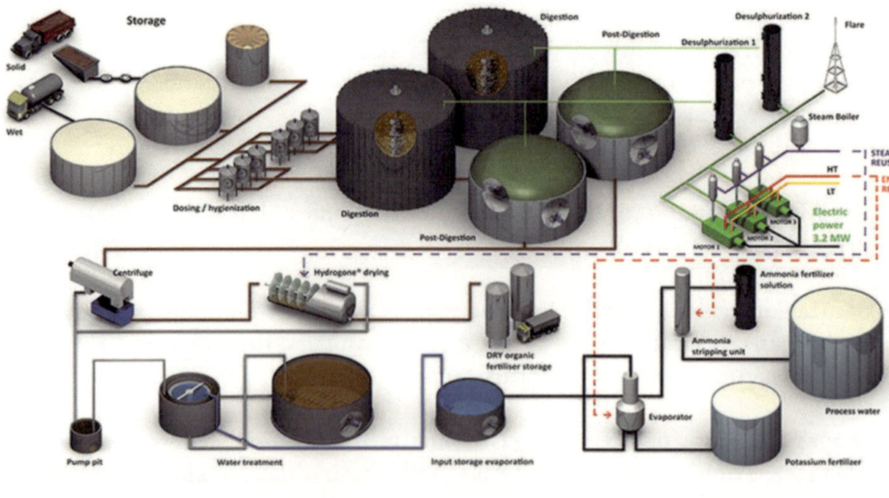

Figure 2.
Anaerobic biodigestion plant (source: BIOTIM®).

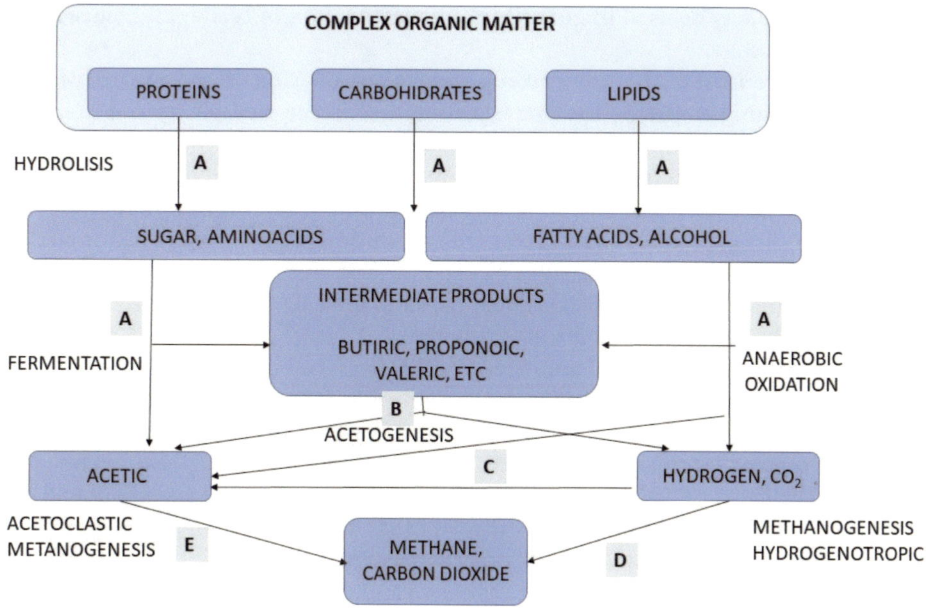

Figure 3.
Stages of the anaerobic biodigestion process [11].

(propionic, butyric, valeric, lactic and ethanol mainly) that have to be oxidized by acetogenic bacteria in the next stage of the process.

Acetogenic stage. While some fermentation products can be directly metabolized by methanogenic organisms (H_2 and acetic), others (ethanol, volatile fatty acids and some aromatic compounds) must be transformed into simpler products, such as acetate ($CH_3\ COO^-$) and hydrogen (H_2), through acetogenic bacteria.

At this stage of the process, most anaerobic bacteria have extracted all the food from the biomass and, as a result of their metabolism, eliminate their own waste products from their cells. These products, simple volatile acids, are the ones that will use as a substrate the methanogenic bacteria in the next stage.

Methanogenic stage. At this stage, a broad group of strict anaerobic bacteria acts on the products resulting from the previous stages. Methanogenic microorganisms can be considered as the most important within the consortium of anaerobic microorganisms, since they are responsible for the formation of methane and the elimination of the medium from the products of the previous groups, being, in addition, those that give name to the general biomethanization process.

Characterization of raw materials. To ensure a good yield in the production of biogas, it is necessary to make a correct selection of the raw materials that are going to be introduced into the biodigester.

For this, it is necessary to carry out previous tests (at laboratory scale), during which the suitability of each of the materials in terms of their methane production potential will be determined.

Biochemical oxygen demand. Biochemical oxygen demand (BOD) provides a measure of biodegradable organics present in a sludge and, in turn, can be used as a metric for the overall effectiveness of an anaerobic digester.

BOD is defined as the amount of oxygen, divided by the volume of the system, taken up through the respiratory activity of microorganisms growing on the organic compounds present in the sample (e.g. water or sludge) when incubated at a specified temperature (usually 20°C) for a fixed period (usually 5 days, BOD5) [12].

This parameter may be used to quantify the concentration of biodegradable organics present in sludge.

The measurement of this parameter provides information on the amount of biodegradable organic matter by the bacteria contained in the biodigester and therefore on the amount of methane that can be produced.

Chemical oxygen demand (COD). Chemical oxygen demand (COD) indicates the amount of oxygen needed to degrade a certain amount of organic matter. In COD tests, a sludge is refluxed in excess with a solution of potassium dichromate and sulphuric acid. As COD measures all organics in a sludge, its value is understandably higher than that of BOD. Thus, the ratio of BOD to COD can be used to represent the biodegradable fraction of a sludge.

Amount of dry matter and volatile solids. Dry matter or total solids are a parameter which indicates the concentration of organic matter that contributed in the feeding of the biodigester. On the other hand, volatile solids inform about the amount of organic matter that ours contains. Depending on the dry matter contained in the digester's input material, the type of digestion to be carried out will be determined, and therefore the type of digester that will be necessary to use for the process. Thus, in those cases where the dry matter concentration is less than 20% of the total, we will be talking about wet digestion and dry digestion when the amount of dry matter is greater than 20% of the total.

Carbon and nitrogen ratio. This is a determining parameter when evaluating the viability and potential performance of an anaerobic digestion process, so it must be considered during the selection process of the raw materials that will be used as inputs of the process. Thus C/N concentrations of 20/1 or 30/1 will give optimal results in digestion [13, 14]. However, low proportions of this ratio may be indicative of high concentrations of ammonium that are detrimental to the process, given the inhibitory nature of the nitrogen compounds process.

Process conditions. In addition to the type and characteristics of the raw material, the state of the same and its composition, the environmental conditions of the process inside the biodigester will largely determine the correct development of the process and the biogas production yield.

Temperature and pH, which are key parameters. Temperature (T) of the process will depend on the type of bacteria used in the process. Depending on the bacteria and the temperature conditions, we can distinguish between different types of processes:

- Mesophilic processes, which take place in a temperature range between 25 and 45°C, typically 35°C.

- Thermophilic processes are those that take place in temperature conditions above 45°C, typically 55°C.

- Psychrophilic processes are the least developed so far, in which the work of bacterias at temperatures around 20°C. Normally the psychrophilic conditions are referred for systems without any heat supply.

As for pH, in ideal conditions it must be in terms of neutrality, avoiding acidic and basic conditions. Excessive acidity (much less than 7) or a medium that is too caustic (much greater than 7) can reduce the metabolism of bacteria or even kill them.

Hydraulic retention time. The hydraulic retention time (HRT). It is the time that passes a flow in the digester from entering until leaving it. It can be referred to the liquid flow rate, the MS it contains or the time that the biomass (bacteria) resides inside the reactor before leaving with the sludge. HRT is an important operational parameter for the anaerobic reactors which can affect the conversion of volatile solids (VS) into biogas. HRT is one of the main parameters to consider when designing the size of the digester [15], and in turn this will depend on the type and conditions of the raw materials used as inputs in the process.

Upgrading from biogas to biomethane or renewable natural gas. It is known as biogas upgrading, the treatment process through which the biogas obtained through anaerobic digestion is transformed into a renewable gas with a quality grade like fossil natural gas [16].

For the biogas to be transformed into renewable natural gas thus increasing its specific caloric value, it is necessary that a series of components such as CO_2, H_2O, sulphur compounds (H_2S), siloxanes, NH_4, O_2 and N_2 be removed from the initial gas stream [17]. For this there are currently different technologies, among which the following stand out. These technologies can be classified in two principal groups:

- Sorption technologies: adsorption and absorption
- Separation: membranes and cryogenic

Pressure swing adsorption (PSA). In this case, carbon dioxide is separated from the biogas by adsorption on a surface under elevated pressure. Molecular sieve materials such as zeolites and activated carbon are commonly used as adsorptive materials for biogas upgrading. The adsorbing material is regenerated by a sequential decrease in pressure before the column is reloaded again, hence the name of the technique.

Some of its advantages are the compactness of the equipment, low energy requirements, low capital investment and simplicity [18]. On the other hand, if hydrogen sulphide is present in the raw gas, it will be irreversibly adsorbed on the adsorbing material. In addition, water present in the raw gas can destroy the structure of the material. Therefore, hydrogen sulphide and water need to be removed before the PSA column.

Physical scrubbing (absorption). In an upgrading plant using the absorption technique, the raw biogas meets a counter flow of liquid in a column which is filled with plastic packing (in order to increase the contact area between the gas and the liquid phase). The principal behind the absorption technique is that carbon dioxide is more soluble than methane. The liquid leaving the column will thus contain increased concentration of carbon dioxide, while the gas leaving the column will have an increased concentration of methane.

Typically, either water or organic solvent (e.g. methanol, N-methyl pyrrolidone and polyethylene glycol ethers) are used to absorb CO_2 in physical absorption plants, whereas amine scrubbing is widely used for chemical absorption. Nowadays, water scrubbing accounts for ~41% of the biogas upgrading market [19].

Water scrubbing. The driving force of the process is the difference of solubility for CH_4 and CO_2 in water. CO_2 is more soluble in water than CH_4 [20]. With the increase in process pressure, this difference becomes higher, and the CO_2 is absorbed more quickly and to a larger quantity. In the scrubber column carbon dioxide is dissolved in the water, while the methane concentration in the gas phase increases. The gas leaving the scrubber has therefore an increased concentration of methane. The water leaving the absorption column is transferred to a flash tank where the dissolved gas, which contains some methane but mainly carbon dioxide, is released and transferred back to the raw gas inlet.

Membrane separation. It is based on the selective permeability property of membranes [21]. Dry membranes for biogas upgrading are made of materials that are permeable to carbon dioxide, water and ammonia. Hydrogen sulphide and oxygen permeate through the membrane to some extent, while nitrogen and methane only pass to a very low extent. Usually membranes are in the form of hollow fibres bundled together. The process is often performed in two stages. Before the gas enters the hollow fibres, it passes through a filter that retains water and oil droplets and aerosols, which would otherwise negatively affect the membrane performance. Additionally, hydrogen sulphide is usually removed by cleaning with activated carbon before the membrane [22].

On the market, three types of membranes are typically used: polymeric, inorganic and mixed matrix membranes. Inorganic membranes have several advantages compared to polymeric, mainly due to their higher mechanical strength, chemical resistance and thermal stability. The current trend in industrial applications is to use mixed matrix membranes [17].

Cryogenic separation. This technique involves subjecting the biogas to high pressures and low temperatures, so that the CO_2 goes into its liquid state, while the methane remains in the gaseous state. Prior to subjecting the raw biogas to the cryogenization process, it must be pretreated to remove the sulphur compounds (H2S) contained "in the gas", since H_2S could damage heat exchangers [23]. On the other hand, volatile organic compounds (VOC) and siloxanes are efficiently removed during the cooling and condensation process, which is a natural part of the cryogenic improvement process.

2.3 Hydrogen

Hydrogen is currently one of the energy vectors that have the greatest potential for medium-term application, within the range of new alternative fuels to conventional petroleum products. The most important characteristic of hydrogen as an energy vector is that the energy yield of hydrogen is about 122 kJ/g, which is 2.75 times greater than hydrocarbon fuels [24].

Hydrogen is a clean fuel without toxic emissions and can easily be applied in fuel cells for electricity generation [25]. It may be stored as a gas under pressure or in a liquid state, or distributed through natural gas networks, thus representing an alternative with a high potential for long-term replacement of natural gas and as a gas compatible with natural gas in the short and medium terms [26].

Although hydrocarbons and coal are currently the main feedstock used for H_2 production [27], the need to increase the integration of renewable technologies will become unavoidable. Thus, this section will focus on the development of hydrogen production technologies compatible with renewable sources.

2.3.1 Hydrogen production

The hydrogen production will depend on two main variables: the feedstock (water or organic materials such as natural gas, biomass, coal or oil) and the origin of the energy to be used in the production process (conventional or renewable energy). If a water disponibility is possible, the main cost of hydrogen production by electrolysis is an electricity cost.

Thermolysis. It involves the application of heat (high temperatures) that causes the water molecule to rupture, resulting in hydrogen and oxygen. The direct decomposition of the water molecule, by thermal processes, requires temperatures around 2.500°K [28]. To reach these temperatures, it is necessary to use solar concentration technologies. However, the application of solar concentrators, to produce hydrogen by direct water thermolysis, leads to material problems and an increase in losses due to re-radiation, reducing absorption efficiency. An effective technique of separation of hydrogen and oxygen is necessary, to avoid an explosive mixture. These handicaps are the reason why there is currently no pilot plant in which the direct decomposition of water takes place, which makes it unfeasible today [29].

As an alternative technique to direct thermolysis, solar thermochemical technology is proposed, based on the reduction of metal oxides [30]. These processes occur in two phases. The first phase consists in the reduction by means of the application of solar energy of the metallic oxide. The second phase, in which the application of heat is not necessary, consists in the exothermic hydrolysis of water, accompanied by the oxidation of the metal, to form hydrogen and the corresponding metal oxide [31].

Electrolysis. Electrolysis involves the breakdown of the water molecule through the application of electrical energy, resulting in hydrogen and oxygen. Traditionally, water electrolysis has been carried out in electrolytic cells formed by two electrodes immersed in an aqueous solution (electrolyte) [32]. Through the application of direct current, the dissociation of the water molecule occurred. The chemical reaction produced during the water electrolysis process is shown below:

$$H_2O \rightarrow 1/2\ O\ 2 + 2H^+ + 2e^-$$

$$2H^+ + 2e^- \rightarrow H_2$$

As already mentioned, the electrolysis process occurs in systems called electrolysers. Next, the main existing electrolyzer technologies are descriptively presented.

Polymeric membrane electrolyzers (PEM). In a polymer electrolyte membrane (PEM) electrolyzer, the electrolyte is a special solid plastic material. Water reacts at the anode to form oxygen and positively charged hydrogen ions (protons). Electrons flow through an external circuit, and hydrogen ions selectively move through the EMP to the cathode. In the cathode, hydrogen ions combine with the electrons in the external circuit to form gaseous hydrogen [33] (**Figure 4**).

Polymeric membrane electrifiers (PEM) are formed by a polymeric membrane, generally manufactured by a commercial polymer perfluorosulfonated acid (brand name of Nafion®) and by two carbon electrodes attached to both sides of the membrane (MEA) [33].

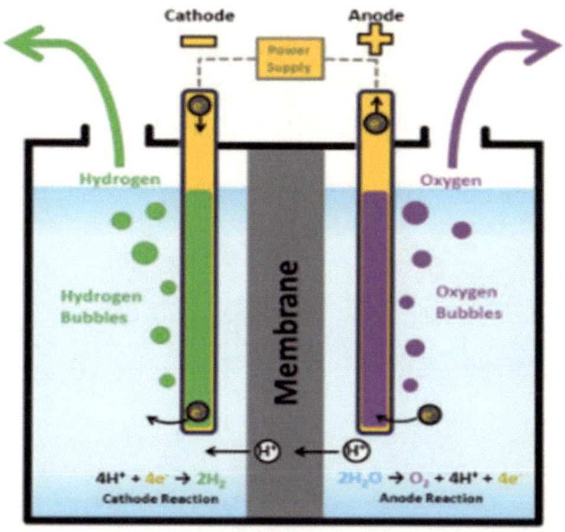

Figure 4.
Scheme of a polymeric membrane electrolyzer (PEM). Source: Green Power Co. Ltd.

Due to its acidic nature, expensive noble metal catalysts such as platinum at the hydrogen and iridium-oxide at the oxygen evolving electrode must be used. Catalyst substances are applied on the electrodes, which favours the processes of electron transfer and therefore increasing the yield of hydrogen production reactions [34].

The MEAs are installed between the so-called bipolar plates. These bipolar plates are responsible for the conduction of water inside the electrolyzer as well as the transport of the gases generated to the exit. In addition, they are responsible for providing mechanical stability to the system, ensuring electrical conductivity and evacuating the heat generated [33].

Alkaline electrolyzers. Alkaline electrolyzers are based on classical technology in which two electrodes are submerged in a solution or electrolyte. In the case of alkaline electrolyzers, the electrolyte consists of a solution of potassium hydroxide (KOH), in concentrations of 30–35%. In these electrolyzers, normally both the electrodes and the tank are made of nickel or a steel alloy with chrome and nickel, materials that can withstand the corrosive power of alkali solutions [35].

In these electrolyzers each of the cells is separated by a diaphragm that prevents the mixing of hydrogen and oxygen produced. These diaphragms are made of ceramic oxides coated with a metallic material. In addition to the membrane, the electrolyzers are formed by a current and voltage adjustment system, a supply water deionization unit and a gas separation device [36] (**Figure 5**).

Alkaline electrolyzers are more economical and durable, are recognized for their mature technology, are safe and can work with pressures up to 25 bars. It is the most mature technology, and due to its development, it is the most economical. Its effectiveness is around 62–82%, being it higher when more larger is the size of the electrolyzer [37].

Chemical conversion processes: reforming and gasification. Another of the technologies for the generation of hydrogen are those based on the processing of organic materials through two technologies called reforming and gasification.

Reforming. The reforming processes are the most common today for obtaining hydrogen. The reforming technology consists of the combination of a combination of hydrocarbons (natural gas, LPG, liquid hydrocarbons) and alcohols with water vapour, resulting in the generation of hydrogens from various chemical reactions [38].

Water steam reforming. Water reforming is based on the reaction of a fuel with water vapour on a catalyst [39]. The water vapour reforming process (known by the acronym SMR, steam methane reforming) may be applied to a wide variety of hydrocarbons (natural gas, LPG, liquid hydrocarbons, etc.) and alcohols. Of all of them, the most used for its availability and ease of handling is natural gas.

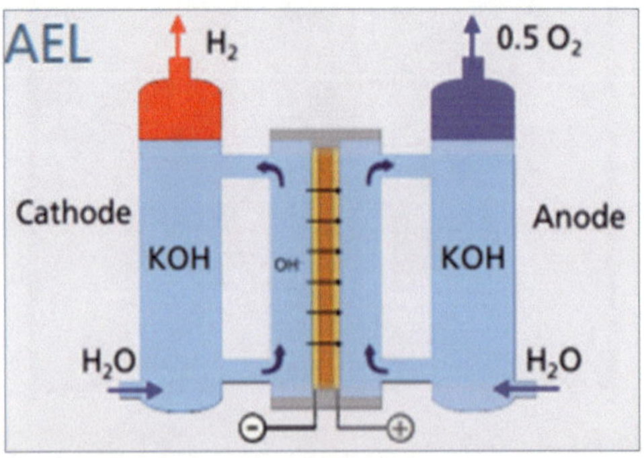

Figure 5.
Scheme of an alkaline electrolyzer. Source: Smolinka-Fraunhofer Institute für Solare Enegiesysteme ISE.

$$CH_4 + H_2O \rightarrow CO + 3H_2$$

This reaction that is verified in the first phase is that of reforming itself, taking place at temperatures around 900°C in tubes through which methane and water vapour circulate through nickel-based catalyst beds. At the end of the process, the gas is directed towards a CO displacement unit in which the following reaction on copper catalysts is verified.

$$CO + H_2O \rightarrow CO_2 + H_2$$

The gas produced as a result of the two previous reactions passes through a condenser in which the water vapour is removed and finally reaches the third stage of the process, the purification process. The gas that reaches this unit is rich in hydrogen with carbon dioxide, water debris, carbon monoxide and methane. This gaseous stream is purified in a membrane separator or adsorption-desorption system (PSA: pressure swing adsorption.) From which, hydrogen is obtained with a purity of 99.999% [38].

Gasification. This technology produces synthesis gas ($CO + H_2$) by controlled heating of an organic waste at temperatures between 800 and 1000°C in an O_2 atmosphere or H_2O vapour. The synthesis gas obtained can be used as a direct fuel, as a source of H_2 or as a chemical raw material to prepare gasoline or diesel by means of the Fischer-Tropsch process. The use of water vapour in the feed allows to increase the production of hydrogen by reducing the production of tars and CO. Gasification, given the severity of the treatment, is particularly indicated for the treatment of plant residues that are difficult to use in other ways.

Among all the reactions involved, the steam reforming and the water-gas reaction should be noted for their importance in determining the degree of gasification and composition of the gaseous products obtained.

$$C_n H_m O_y + (2n - y) H_2O \rightarrow nC0_2 + (2n - y + m/2) H_2$$
$$\text{(reforming or hydrolysis reaction)}$$

$$CO + H_2O \leftrightarrow CO_2 + H_2 \text{ (water-gas equilibrium reaction)}$$

Fermentative hydrogen. Biological hydrogen production is possible under anaerobic conditions [40]. Some enzymes, called hydrogenases, are able to create H_2 molecules from H+, which is released as consequence of the anaerobic degradation of simple organic molecules, as sugars, into small fatty acids (mainly acetic acid). Digesters, with the same configuration as that of the ones used for biomethane production, can produce H_2 from different organic wastes. Shorter hydraulic retention times must be applied in order to avoid H_2 consumption by other groups of organism.

Photosynthetic hydrogen. Some species of microalgae contain hydrogenases that transform H^+ to H_2 under the absence of oxygen and solar illumination [41]. This process is a consequence of continuous electron motion in the light-related reactions. These electrons are used to reduce H^+ resulting in the creation of H_2 molecules.

2.4 Distribution and supply: biomethane, hydrogen, and natural gas

One of the main challenges that currently arise, to make viable the introduction of alternative fuels for vehicular use, is the logistics and supply of these fuels, from origin to the different points of supply.

Currently the only gas (not derived from petroleum), marketed for vehicular use, is natural gas. The supply of vehicular natural gas (NGV) is supplied through service stations or gas stations, to which natural gas arrives through road transport (in cryogenic or pressure tanks) and through the connection to networks of gas transportation.

The introduction of other alternative gases such as renewable natural gas (biomethane) and hydrogen is posing important challenges for gas companies, since the characteristics of these gases, although compatible with pre-existing natural gas networks, present a series of differences that will make the adaptation of the current infrastructure necessary and the need to create new distribution and distributed generation models.

Biomethane. Biogas generated in anaerobic digestion plants from agro-food waste, sewage treatment plants, landfills, etc. once purified and transformed into biomethane, with a quality equivalent to natural gas, might be injected into natural gas grids. For this, it will be necessary that the continuous measurement equipment is available at the injection point, which certifies the quality of the gas. There is a European regulation, UNE-EN 16726 Gas infrastructure, to standardize gas quality and the main parameters and their limits for the gases to be transported and/or injected into the natural gas network.

The control by the managers of the natural gas distribution networks and of the distribution of this new gas and the necessary quality and safety problems, to ensure that the distributed gas maintains the necessary conditions, implies an important technological update, which will require the implementation of new management and control tools, from the injection points to the supply.

To keep track of this valuable characteristic after having been mixed with fossil natural gas, a tracking mechanism is needed. Mass or energy balances serve reliable and complete retracing of biomethane from its production site to the final consumer.

Biomethane trade predominantly takes place in the country of its production. There are only a few examples of physical cross-border biomethane trade, e.g. from Germany to Sweden and to the Netherlands as well as from the UK to the Netherlands. If the market is not balanced, meaning demand exceeds supply or vice versa, cross-border trade is able to increase flexibility and transfer biomethane where it is needed. Barriers are often created by the different national regulatory frameworks, but they can be removed by harmonizing the national tracking systems which means that two different biogas registers are able to exchange biomethane amounts from the country of production to the country of final consumption.

Hydrogen. Being considered an energy vector this is a compatible gas with natural gas networks. However, this compatibility does not mean that H_2 can be transported as such in the same network; since it is a different molecule, it will be necessary to look for ways to be combined with natural gas or biomethane, obtaining a final gas with greater energy potential.

Thus, two main formulas are planned: injection of H_2 into the network (P2G), both in different % of H_2 injected into natural gas (power to gas—H_2), and the methane of H_2 for injection as synthetic natural gas.

The production of hydrogen, from the surplus of electricity, is an optimal way to absorb the excess (renewable) energy produced at times of demand, as an alternative form of storage [42]. Thus the energy 'stored' in the generated hydrogen (acting as an energy vector) can be combined with natural gas or biomethane transported in the gas network.

Alternatively, the H_2 produced can be recovered through CO_2 methane (discarded in the biogas to biomethane upgrade process), resulting in methane of renewable origin.

In economic terms, if connected to the grid, hydrogen can be produced subject to short-timescale variations in the power market or under flat rates through power purchase agreement (PPA) contracts. In the first case, production will happen especially at moments of low- and medium-power prices. There will be a certain number of operating hours at higher prices, causing hydrogen costs to increase. Electrolyzers may be operated as demand response assets to support energy balance over the grid [43], making profits from balancing or ancillary services (IRENA, 2018c). In the second case, hydrogen is produced at flat power rates under PPA contracts. In this case, operation can be continuous, which can improve the overall efficiency of the process—the higher the number of operating hours, the lower the hydrogen production costs. However, such baseload operation reduces the flexibility of the power system.

Other options to transport renewable fuels. For those cases where the transport of large volumes of gas, whether biomethane or hydrogen, is not required, they can be transported as compressed or liquefied gases in the cryogenic state. The choice of another method will depend on the amount of gas as well as the destination of the gas. Transport in the gaseous state does not allow the accumulation of large volumes of gas, unlike liquefaction (high pressure and low temperature) that will allow the storage of larger volumes.

Both liquefaction and gas compression will lead to important energy requirements, which will influence the final cost of gas distribution. The decision on the form of gas transportation will be linked to the other factors that affect the gas logistics chain, including the form of production (centralized or distributed), supply chain as well as access to the point of use.

As for transport methods, renewable gases, like other gases, can be distributed by ship, train and tank trucks. However, contrary to what happens with other gases of non-renewable origin such as natural gas, one of the main advantages of renewable gases is that these can be generated at points close to those of use, avoiding the need to carry out transportation on large routes, which involves the intensive use of energy.

2.4.1 Distributed production

As an alternative to the injection of biomethane and hydrogen in the natural gas distribution network, an alternative form of supply to end users is proposed, avoiding the use of the gas network. In those cases, in which the green gas generation location (biomethane or hydrogen) is far from the gas network, an isolated model or diffuse model is proposed.

Regarding biomethane as an alternative to the large biogas production facilities, linked to centralized biogas agro-livestock plants, or sewage treatment plants or urban waste plants, a model that has gained strength in recent years is the implantation of small-/medium-size biogas production plants with by-products of agricultural origin. These installations are mostly located in places far from large population centres and therefore without the possibility of connection to the gas network. To solve this handicap, these biogas production plants (and biomethane), if they can be close to communication roads (roads), with a large transit of livestock vehicles (tractors and other machinery) and trucks. Therefore, the production of biomethane in these plants, and the installation of dispensing stations for this renewable gas (gas stations), could be an important alternative to the transport of renewable gas through the gas network.

An alternative solution for isolated hydrogen generation and distributed supply would be through the development and implementation of so-called hydrogenators, or hydrogen production stations from renewable energy (photovoltaic and/or wind), which is stored in batteries being used on demand for the production of hydrogen through the use of electrolyzers. In this way, this new fuel could be given access, in areas or road networks far from the main population centres.

3. Discussion and conclusion

3.1 Biogas and renewable natural gas (biomethane)

In recent years there has been a great controversy between detractors and defenders in relation to the environmental benefits of natural gas in general and renewable natural gas, against the use of conventional fossil fuels (diesel and gasoline) for vehicular use.

Among the detractors, various environmental groups and associations are used as main arguments against the benefits attributed to the use of gas as an alternative fuel, in that combustion of this in engines generates similar concentrations and even higher pollutant gases than fossil fuels. On the contrary, the studies that defend the environmental advantages of the use of gas as fuel defend the low emissions (mainly of particles) of the gas against other fossil fuels.

In this chapter, different alternative fuels to conventional fuels have been highlighted. In this sense, although it is true that CO_2 emissions from both natural gas and renewable natural gas will be very similar to those produced by conventional fuels, since both types of fuels have an organic (carbon) origin, it is important to highlight the other environmental advantages that alternative gas fuels have compared to conventional ones.

As already mentioned, the intention of presenting natural gas as a fuel has not been so much for its direct environmental benefits (which as it has already been described in a certain way if it has them) but for the importance that this can mean as a bridge to greater use of renewable natural gas.

Some of the environmental (and socio-economic) advantages that the deployment of renewable natural gas can present in a future scenario in which other energy sources such as electricity coexist could be the following.

3.1.1 Neutral CO_2 balance

However, as already said, the combustion of renewable natural gas (and biogas), if it would generate CO_2 emissions, given the renewable origin of the gas, these emissions would count as neutral in a global balance, given that the origin of the biogas is from organic materials, which in their origin have already absorbed atmospheric CO_2.

In addition to this advantage (neutrality), the management of organic waste for the production of biogas, if carried out in a controlled manner, avoids the bad or nonmanagement of organic waste (household waste, animal waste, other organic waste) that would entail the emission of CH_4 (uncontrolled natural digestion), with the important effect that this gas supposes in terms of atmospheric heating.

3.1.2 Atmospheric emissions

Although it is true that the use of biogas and biomethane as alternative fuels will not have a great impact on the global balance of CO_2 emissions if it will be a significant reduction in other emissions such as sulphur compounds and the suspended particles.

Specifically, it has been shown that in heavy vehicles, natural gas reduces NOx emissions up to 86% and particles 75%. In the case of buses, natural gas reduces NOx emissions up to 90% and particle emissions up to 69% compared to diesel. Additionally, natural gas does not have sulphur in its composition, so unlike diesel, it eliminates SOx emissions (Foundation for the Promotion of Industrial Innovation of the Higher Technical School of Industrial Engineers of the Polytechnic University of Madrid).

3.1.3 Circular economy and local development

In addition to the direct environmental advantages, it is necessary to highlight other advantages implicit in the use of biogas and biomethane, such as those linked to the new circular economy scenario and the much needed revitalization of rural areas.

The management of organic waste for the production of biogas for automotive and transportation allows to contribute to the so-called circular economy, whose purpose is the use of all waste generated, so that these are transformed into resources and thus close the productive circle, avoiding the demand for new resources and wasting on raw materials useful for other uses.

As for the different formats of production and management of renewable natural gas, as mentioned in the previous sections, a possible scenario for the development of an economy based on the use of waste and by-products for the generation of biogas could be based in a distributed generation model, in which the biogas generation points are directly linked to the production centres of the by-products. In this way, a broad framework of development possibilities could be opened at the local level, which would allow establishing a development path for those rural areas most in need of alternatives.

3.2 Hydrogen

It is presented in a very different scenario from that of biogas, since both its origin and its use as an alternative fuel are presented in a scenario with large differences compared to other alternative gases.

Hydrogen, as explained in the previous sections, originates from different processes that involve the use of external energy to cause the dissociation of the water molecule, releasing the hydrogen atoms. To carry out this process, depending on the origin of the energy source, it may be considered an environmentally sustainable process or not. In the case of hydrogen production, from renewable electricity (photovoltaic, wind or hydroelectric solar), the hydrogen thus produced if it can be considered a renewable fuel with zero emissions.

Hydrogen, in its use as a fuel in transport, is generally used to feed fuel cells, which generate electricity that is supplied to an electric motor, obtaining as its only residual gas water vapour, which is a completely free energy source of emissions. Also, hydrogen, in combination with residual CO_2 (e.g. from biogas upgrading processes), through a process known as power to gas, would lead to the production of methane that can be used, for example, in vehicles powered by renewable natural gas.

Additionally, neither atmospheric emissions nor CO_2 emissions are a problem for the use of hydrogen. In addition, the great advantage of using this technology, since as the only exhaust gas, water vapor is obtained, from the use of hydrogen as an energy source. With this, the cycle would be closed because hydrogen produced with water is returned to the environment.

Author details

Carlos Repáraz Martín[1*], Ignacio de Godos Crespo[2], Marcelo F. Ortega Romero[1] and Bernardo Llamas Moya[1]

1 Polytechnic University of Madrid, ETSMI, Madrid, Spain

2 University of Valladolid, EIFAB, Soria, Spain

*Address all correspondence to: carlos.reparaz.martin@alumnos.upm.es

References

[1] European Commission. EC: 2009/28/EC. DIRECTIVE (EU) on the promotion of the use of energy from renewable sources and amending and subsequently repealing Directives 2001/77/EC and 2003/30/EC. 2009. Available at: https://eur-lex.europa.eu/eli/dir/2009/28/oj

[2] Faramawy S, Zaki T, Sakr AA-E. Natural gas origin, composition and processing: A review. Journal of Natural Gas Science and Engineering. 2016;**34**:34-54

[3] Speight JG. Natural Gas: A Basic Handbook. Houston, Texas: Gulf Publishing Company; 2007

[4] Major A. "Eternal flames": Suicide, sinfulness and insanity in "western" constructions of sati, 1500-1830. International Journal of Asian Studies. 2004;**1**(2):247-276. DOI: 10.1017/S1479591404000233

[5] Kidnay AJ, Parrish WR, McCartney DG. Fundamentals of Natural Gas Processing. 2nd ed. CRC Press, Taylor and Francis Group, LLC.; 2011

[6] Lloyd R, Snowdon. Natural gas composition in a geological environment and the implications for the processes of generation and preservation. Organic Geochemistry. 2001;**32**(7):913-931. DOI: 10.1016/S0146-6380(01)00051-1

[7] BP Energy Outlook. London; 2019

[8] Strauch S, Krassowski J, Singhal A. Biomethane Guide for Decision Makers–Policy Guide on Biogas Injection into the Natural Gas Grid. Fraunhofer; 2013

[9] Castells XE. Waste energy treatment and recovery. In: De Santo D, editor. Tratamiento Y Valoracion Energetica De Residuos. 2005. ISBN: 9788479786946

[10] Zhao Q, Leonhardt E. Purification technologies for biogas generated by anaerobic digestion. Compressed biomethane. CSANR Research. 2010, 2010;**1**:21-38

[11] Pavlostathis SG, Giraldo-Gomez E. Kinetics of anaerobic treatment: A critical review. Journal Critical Reviews in Environmental Control. 2009;**21**:411-490

[12] Jouanneau S, Recoules L, Durand MJ, Boukabache A, Picot V, Primault Y, et al. Methods for assessing biochemical oxygen demand (BOD): A review. Water Research. 2014;**49**:62-82

[13] Varnero Moreno MT. Biogas manual. In: Removal of Barriers for Rural Electrification with Renewable Energies; Project CHI/00/G32. Chile: FAO; 2011

[14] Al Seadi T. Biogas from AD: BIOEXCELL Training Manual. Project Deliverable of the BIOEXELL Project. Biogas Centre of Excellence; 2004

[15] Dareioti MA, Kornaros M. Effect of hydraulic retention time (HRT) on the anaerobic co-digestion of agro-industrial wastes in a two-stage CSTR system. Bioresource Technology. 2014;**167**:407-415

[16] Bortoluzzi G, Gatti M, Sogni A, Consonni S. Biomethane production from agricultural resources in the Italian scenario: techno-economic analysis of water wash. Chemical Engineering. 2014;**37**

[17] Prussi M, Padella M, Conton M, Postma ED, Lonza L. Review of technologies for biomethane production and assessment of Eu transport share in 2030. Journal of Cleaner Production. 2019;**222**:565-572

[18] Delgrado JA, Uguina MA, Sotelo J, Ruiz B, Gomez J. Fixed bed adsorption of carbon dioxide/methane mixtures

on silicalite pellets. Adsorption. 2006;**12**:5-18

[19] Toledo-Cervantes A, Estrada JM, Lebrero R, Muñoz R. A comparative analysis of biogás upgrading technologies: Photosynthetic vs physical/chemical processes. Algal Research. 2017;**25**:237-243

[20] Kapoor R, Subbarao PMV, Vijay VK, Shah G, Shota S. Factor affecting methane loss from a water scrubbing based biogas upgrading system. Applied Energy. 2017;**208**:1379-1388

[21] Ryckebosch E, Drouillon M, Vervaeren H. Techniques for transformation of biogas to biomethane. Biomass and Bioenergy. 2011;**35**:1633-1645

[22] Petersson A, Wellinger A. Biogas upgrading technologies developments and innovations. In: IEA Bioenergy. 2009

[23] Hoyer K, Hulteberg C, Svensson M, Jernberg J, Norregaard Ø. Biogas upgrading—Technical Review. 2016

[24] Kapdan IK, Kargi F. Bio-hydrogen production from waste materials. Enzyme and Microbial Technology. 2006;**38**:569-582

[25] Hosseini SE, Wahid MA. Hydrogen production from renewable and sustainable energy resources: Promising green energy carrier for clean development. Renewable and Sustainable Energy Reviews. 2016;**57**:850-866

[26] Eveloy V, Gebreegziabher T. A review of projected power-to-gas deployment scenarios. Energies. 2018;**11**(7):1824. DOI: 10.3390/en11071824

[27] Mueller-Langer F, Tzimas E, Kaltschmitt M, Peteves S. Techno-economic assessment of hydrogen production processes for the hydrogen economy for the short and medium term. International Journal of Hydrogen Energy. 2007;**32**(16):3797-3810. DOI: 10.1016/j.ijhydene.2007.05.027

[28] Baykara SZ. Experimental solar water thermolysis. International Journal of Hydrogen Energy. 2004a;**29**(14):1459-1469. DOI: 10.1016/j.ijhydene.2004.02.011

[29] Baykara SZ. Hydrogen production by direct solar thermal decomposition of water, possibilities for improvement of process efficiency. International Journal of Hydrogen Energy. 2004b;**29**(14):1451-1458. DOI: 10.1016/J.IJHYDENE.2004.02.014

[30] Nakamura T. Hydrogen production from water utilizing solar heat at high temperatures. Solar Energy. 1977;**19**(5):467-475

[31] Steinfeld A. Solar thermochemical production of hydrogen—A review. Solar Energy. 2005;**78**(5):603-615. DOI: 10.1016/J.SOLENER.2003.12.012

[32] Holladay JD, Hu J, King DL, Wang Y. An overview of hydrogen production technologies. Catalysis Today. 2009;**139**(4):244-260. DOI: 10.1016/j.cattod.2008.08.039

[33] Grigoriev SA, Porembsky VI, Fateev VN. Pure hydrogen production by PEM electrolysis for hydrogen energy. International Journal of Hydrogen Energy. 2006;**31**(2):171-175. DOI: 10.1016/J.IJHYDENE.2005.04.038

[34] Tijani AS, Ghani MFA, Rahim AHA, Muritala IK, Binti Mazlan FA. Electrochemical characteristics of (PEM) electrolyzer under influence of charge transfer coefficient. International Journal of Hydrogen Energy. 2019;**44**(50):27177-27189. DOI: 10.1016/j.ijhydene.2019.08.188

[35] Zeng K, Zhang D. Recent progress in alkaline water electrolysis for hydrogen production and applications. Progress in Energy and Combustion Science. 2010;**36**(3):307-326. DOI: 10.1016/J.PECS.2009.11.002

[36] LeRoy RL. The thermodynamics of aqueous water electrolysis. Journal of the Electrochemical Society. 1980;**127**(9):1954. DOI: 10.1149/1.2130044

[37] Kothari R, Buddhi D, Sawhney RL. Comparison of environmental and economic aspects of various hydrogen production methods. Renewable and Sustainable Energy Reviews. 2008;**12**(2):553-563. DOI: 10.1016/j.rser.2006.07.012

[38] Chaubey R, Sahu S, James OO, Maity S. A review on development of industrial processes and emerging techniques for production of hydrogen from renewable and sustainable sources. Renewable and Sustainable Energy Reviews. 2013;**23**:443-462. DOI: 10.1016/j.rser.2013.02.019

[39] Díez GE. Bio-Hydrogen Production Through Catalytic Biomass Gasification with Integrated CO_2 Capture. Spain: University of Oviedo; 2017

[40] Ghimire A, Frunzo L, Pirozzi F, Trably E, Escudie R, Lens P, et al. A review on dark fermentative biohydrogen production from organic biomass: Process parameters and use of by-products. Applied Energy. 2015;**144**:73-95

[41] Allakhverdiev SI, Thavasi S, Kreslavski VD, Zharmukhamedov SK, Klimov VV, Ramakrishna S, et al. Photosynthetic hydrogen production. Journal of Photochemistry and Photobiology C: Photochemistry Reviews. 2010;**11**(2-3):101-113

[42] Acar C, Dincer I. Comparative assessment of hydrogen production methods from renewable and non-renewable sources. International Journal of Hydrogen Energy. 2014;**39**(1):1-12. DOI: 10.1016/J.IJHYDENE.2013.10.060

[43] IRENA. Hydrogen a renewable energy. Report prepared for the 2nd Hydrogen Energy Ministerial Meeting in Tokyo, Japan. Irena; 2019

Chapter 7

Natural Gas as a Gateway for Renewable Gas in Transport

Eugenia Sillero Maté, Carla García, Paloma Martínez Ramos and Elena Martínez Calvo

Abstract

Mobility with biomethane is already a reality in many countries in Europe, while in others it is just the beginning. Biomethane is a green and clean fuel which is obtained from the biogas produced by the anaerobic decomposition of the organic matter present in urban waste, sewage water and agricultural, livestock and forest residues. To generate biomethane, the biogas undergoes a process called upgrading. It is a process of elimination of different components such as CO_2, which can be used to obtain synthetic natural gas. Biomethane is perfectly compatible with natural gas; it can be injected in the pipeline grid or used directly as fuel in a vehicle, generating a null CO_2 balance or even negative depending on the type of residue it originated from. Biomethane represents a real solution to decarbonise transport. However, applying this solution depends mainly on public policies and the commitment at national and European level. This chapter will analyse the impulse models for the production of biomethane and its use as fuel for the transport sector. In addition, it will assess whether the development of this market would be possible without public support by creating a product ruled exclusively by the market.

Keywords: mobility, biomethane, decarbonisation, natural gas, green transport

1. Introduction

There are a lot of uncertainties about what decarbonised transport will be like in 2050; however, it is increasingly accepted that clean gas or low-carbon gas and hydrogen will form an essential part of the energy mix, especially in segments where electrification is difficult, such as heavy-duty transport and maritime transport.

At present, natural gas technology is mature and available for all kinds of transport, from light vehicles to vans, buses, lorries, trains and ships. The growth of the fleet of vehicles and the fleet of large ships continues to increase given that it is an alternative that guarantees air quality by significantly reducing the emissions of pollutants that affect health (nitrogen oxides, particulates and sulphur oxides) and helping to reach the targets thanks to a reduction of CO_2 emissions. However, this reduction is insufficient. The solution depends on the progressive incorporation of biomethane, hydrogen and power-to-gas technologies that do not require modification of distribution grids or the engines of users.

Conventional gas faces the major challenge of decarbonisation, and to do achieve this, there are obstacles it must overcome. Green gas, like other renewable energies, is more expensive than fossil fuels; stakeholders are not sufficiently prepared to invest, and, therefore, its development in Europe is taking place with aid policies.

The regulatory framework is another of the major drawbacks. It is obvious that this framework was not designed with the existence of renewable gases in mind and it needs to evolve at the same pace as technology.

The Agency for the Cooperation of Energy Regulators (ACER) recently indicated that "It seems clear that a sustainable future needs decarbonised gases and new technologies (such as P2G), but the current regulatory framework was not designed with these activities in mind and the lack of regulation for these areas may have unintended consequences, acting as a barrier or hindrance to their development" [1].

In this article we shall analyse the incentive schemes for the production and use of biomethane in different European countries. Before doing so, we shall begin with a description of the challenge that the mobility sector must face over the coming years, the obstacles that it must overcome and the contribution of green gas to this end.

2. The challenge of decarbonising transport

The European Union, as a worldwide benchmark in the fight against climate change, has established a roadmap named the Climate and Energy Framework which represents a global strategy comprised of three main targets that involve all sectors of the economy at multiple levels. The targets for 2030 are as follows: 40% cuts in CO_2 emissions, in relation to 1990 levels, a market share of 32% share for renewable energy in final energy consumption and a 32.5% reduction of energy consumption.

With regard to mobility, different standards establish the specific targets for the decarbonisation of transport.

Europe has established a target for the share of renewable energy in transport through Directive (EU) 2018/2001. By 2030, 14% of the energy used in the sector must come from renewable sources, and 3.5% of this figure must be covered by biogas that does not come from crops.

CO_2 emission limits have also been regulated, establishing a deadline that is truly demanding and a long way from actual emissions at present.

The Regulation (EU) 2019/631 of the European Parliament and of the Council of 17 April 2019 setting CO_2 emission performance standards for new passenger cars and for new light commercial vehicles establishes a target to reduce CO_2 emissions for passenger cars by 37.5 and 31% for vans by 2030, in both cases in relation to the levels recorded in 2021. An intermediate target has also been set, according to which passenger cars as well as cans will have to emit 15% less CO_2 in 2025.

That is to say, the average of the emissions from fleets of new vehicles should not exceed 95 grammes of CO_2 per kilometre in 2021, 81 grammes of CO_2 per km in 2025 and 67 grammes of CO_2 per kilometre in 2030.

Some experts believe that the wide gap that exists between actual emissions and targets could result in a cataclysmic disaster for the automotive industry. According to a report by JATO [2], which analyses the emissions of different manufacturers in 2018, the difference is so vast that it could generate penalty payments of up to 34 billion Euros.

Inevitably, new heavy-duty vehicles will also have to considerably reduce their emissions. Regulation (EU) 2019/1242 establishes a 15% reduction of CO_2 emissions for heavy-duty transport by 2025 and 30% by 2030, in comparison with the levels recorded in 2019.

There is even greater complexity in this case than with light vehicles, given that the electrification of heavy-duty vehicles with batteries is an issue that has not yet been resolved and the same applies to maritime transport.

3. Green gas—biomethane, hydrogen and P2G—, a solution

Biomethane is obtained from methane emissions resulting from the anaerobic digestion of organic matter found in urban waste, waste water and agricultural, livestock and forestry waste. To produce biomethane, these emissions from the decomposing organic matter, which would otherwise be emitted into the atmosphere and contribute to the greenhouse effect, are captured and used to produce fuel for vehicles. For this reason, biomethane acts as a sink, and the greenhouse gas emissions of a vehicle powered with this fuel are considered neutral (**Figure 1**).

Another way of decarbonising gas is through hydrogen or synthetic natural gas (power-to-gas).

Hydrogen is set to play an important role in the energy mix of the European Union, and this is reflected in the National Energy and Climate Plans of a significant number of Member States. "Green hydrogen", produced through electrolysis of water using renewable energy, offers high decarbonisation potential when used in a fuel cell in order to power a vehicle or if it is injected into the gas grid. It is widely accepted that at least 10% of hydrogen can be mixed into the natural gas system without the need to make changes to the grid or modifications to the engines of users.

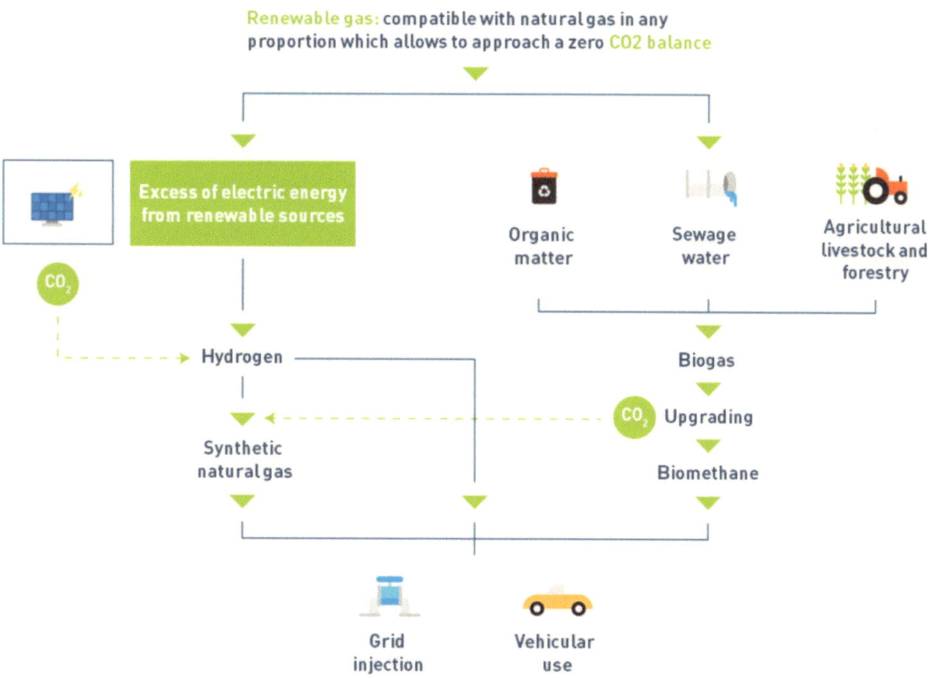

Figure 1.
Process to generate renewable gas.

It is also possible to combine this green hydrogen with the CO_2 extracted and stored by the industrial sector in order to generate synthetic gas which, at a practical level, is the same as natural gas and can be used in vehicles or injected directly into the pipeline grid as a replacement for natural gas.

4. Real analysis of the emissions of a vehicle

When it comes to addressing the sustainability of transport, today's society faces two major challenges;, on one hand, it must guarantee air quality in order to safeguard the health of citizens, and, on the other, it must protect the planet from climate change.

Air quality is determined by the local emissions of nitrogen oxides, sulphur oxides (SOx) and particulate matter (PM). All are local pollutants, harmful to health, and therefore they are measured as they come out of the exhaust.

The same does not apply with the emissions responsible for global warming, which are largely responsible for causing devastating consequences for our planet: an increase in average temperature, melting of ice, rising sea levels and extreme weather events with the damage this entails.

Making an assessment of the impact of climate change on transport while limited to observing the exhaust emissions of a vehicle is totally insufficient; a comparison in fair conditions necessarily requires an analysis of the product life cycle or at least with a well-to-wheel approach.

With regard to the impact of the use of natural gas on air quality, according to a report produced by the Foundation for the Promotion of Industrial Innovation of the Polytechnic University of Madrid [3] which analyses scientific studies that have used proven methods of measuring atmospheric emissions from different types of vehicles, under real conditions of urban and interurban traffic, it is seen that natural gas guarantees air quality by reducing NOx emissions by up to 97% and particulates by up to 70%.

As we mentioned, when we talk about reducing CO_2 emissions, it is not enough to analyse the effluent from the exhaust pipe. The Greenhouse Gas Intensity from Natural Gas in Transport report prepared by Thinkstep [4] assesses emissions from a wide range of fuels with a well-to-wheel approach, that is, considering emissions throughout the process from the generation of energy to its use in the vehicle (**Figure 2**).

In this study we observe that the greenhouse gas emissions of a vehicle powered by biomethane are 83% lower than those of a vehicle powered by gasoline and 79% lower than a diesel vehicle.

Recently, the French Institute of Petroleum (IFP Energies Nouvelles) [5] in France published a study evaluating the emissions of CO_2 in different means of road transport in two time horizons (2019 and 2030) taking into account the life cycle of

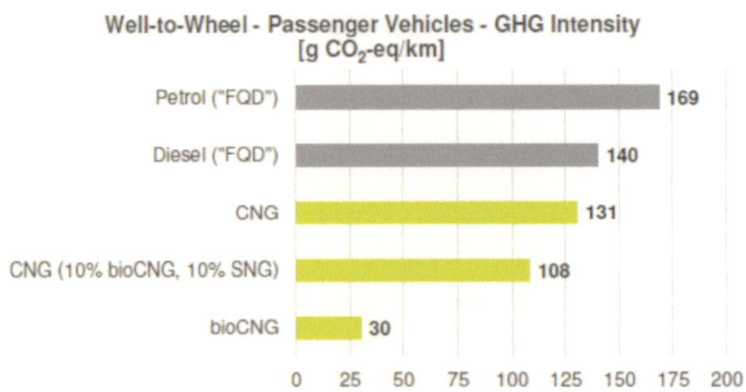

Figure 2.
Comparative analysis of different fuels.

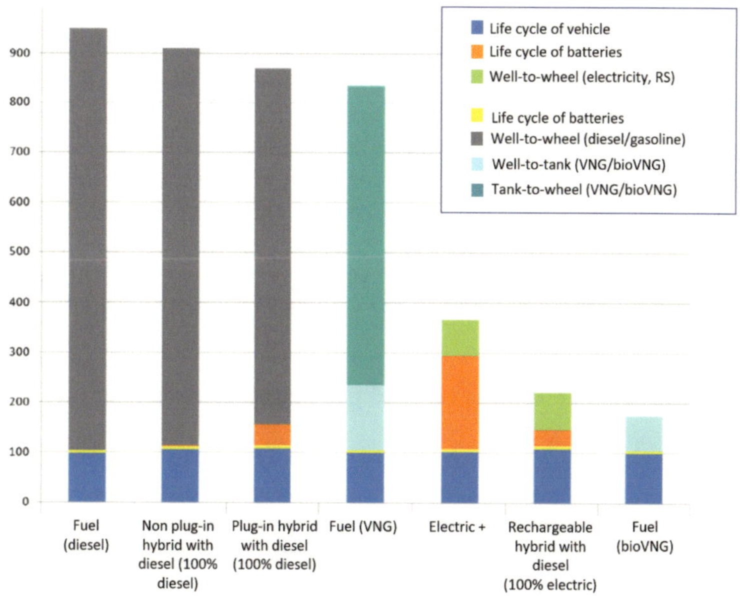

Figure 3.
CO_2 emissions (g/km) in light commercial vehicles in 2019 scenario.

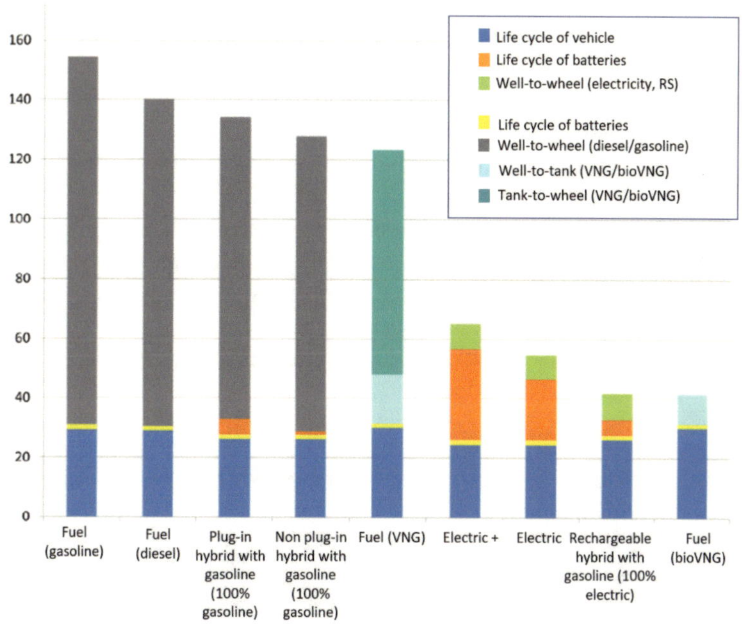

Figure 4.
CO_2 emissions (g/km) in passenger cars in 2019 scenario.

the vehicle and the fuel. The report concludes that, in both light and utility vehicles and even in 12 tonne lorries, the use of biomethane provides the best results in terms of greenhouse gas emissions.

However, electric vehicles with high capacity batteries are penalised by the significant amount of CO_2 emitted during the manufacturing of batteries, largely deriving from the extraction and refining of the metals used such as lithium, cobalt and nickel and due to energy-intensive processes used for the manufacturing and assembly of the cells (**Figures 3** and **4**).

These reports reach a clear conclusion: The European legislative framework should also consider analysis methods that make it possible to establish homogeneous comparisons regarding the environmental impact of different technologies, so that the effect of biomethane can be considered in order to accredit compliance with the environmental targets. However, the regulation published recently to mitigate the effect of transport on climate change only deals with exhaust emissions.

The last regulatory package has made a first step in this direction. The aforementioned Regulation (EU) 2019/1242 outlines, in a review clause for 2022, the development of a methodology that analyses the cycle of fuel from the point of production and not only the direct emissions from the vehicle.

5. Natural gas: vector for the entry of green gas in mobility

At present, natural gas technology is mature and available for the entire range of vehicles, from heavy-duty vehicles to vans, passenger cars and ships. As we have mentioned, it is an alternative that guarantees air quality and reduces CO_2 emissions, making it possible to reach a neutral balance of emissions when renewable gas is used.

In accordance with demanding European regulation, sectors such as heavy-duty transport and maritime transport have mature technology in order to reduce their emissions through the use of natural gas and the progressive incorporation of renewable gas.

In Europe there are already 1,331,000 vehicles on the roads that are powered by natural gas, and the forecast for 2030 is 13 million vehicles. The same applies to the supply infrastructure. At present, in Europe there are 3684 gas stations for compressed natural gas and 228 for liquefied natural gas which will rise to 10,000 in the case of CNG and 2000 in the case of LNG by 2030 [6].

The maritime sector also continues to progress; according to data from DNV, there are currently 170 LNG-fuelled ships in operation worldwide, another 112 are ready to use this fuel, and 35 are currently being built.

The current development and consolidation of the use of natural gas in mobility facilitates the progressive penetration of renewable gas in different means of transport, and it could be an important resource for meeting the demanding targets that the automotive sector faces over the coming years by reducing the foreseeable sanctions.

Below, an analysis has been carried out of the support systems that are being implemented to assist this transition.

6. The main mechanisms through which renewable gas has been incentivised in Europe

In order to give a boost to renewable gas to enable it to offer all of its potential to the decarbonisation process, the different Member States have developed incentive mechanisms that make it possible to create a suitable ecosystem for the development of renewable gas.

Feed-in tariff [7]: Feed-in tariff is an incentive linked to the type of production technology. In this case, the producer ensures the price of the renewable gas for the next 10–20 years, as well as access to the grid. An example of the benefit of these kinds of tariffs is reflected in the pricing structure for the purchase of renewable gas in France. This price depends on the characteristics of the production plant and promotes sustainable projects by adding premiums according to the origin of the waste. In this case, for 15 years, the producer sells all of their biomethane production to the supplier they choose, thus encouraging the long-term vision that this type of technology offers.

Feed-in premium: Feed-in premium is a premium on top of the market price of the production costs in accordance with the characteristics of the project.

Tax incentives: The most common incentive is the exemption from taxes on emissions or fossil fuels.

Guarantee of origin: This is both a regulatory mechanism and a support mechanism. These certificates document the chain of custody of the renewable gas injected into the natural gas grid. Their objective is to certify the renewable character of the gas and its main characteristics and to provide traceability.

7. European experience with renewable gas

The biogas market is very heterogeneous; it varies greatly from one country to the next. This analysis aims to provide a comparative analysis of the national experiences of different countries, assessing the current situation in the biomethane sector and evaluating possible development trends in terms of mobility.

The analysis was carried out in seven European countries that are benchmarks for the production of biogas given that they represent 90% [9] of total EU production. This analysis is based on a study by KPMG, Regulatory benchmarking of renewable gas in Europe [8], carried out by Gasnam, Sedigas and Aebig.

7.1 France

The support system for renewable gas in France involves a feed-in tariff for the injection of biomethane into the gas grid. The French system also has guarantees of origin.

Natural gas suppliers and biomethane producers establish bilateral contracts for the purchase and sale of renewable gas. Producers inject this biomethane into the grid, after which, the suppliers receive the renewable gas along with a guarantee of origin that accredits the renewable origin of the gas.

There is a maximum period of 1 year to create this guarantee of origin after injecting the biomethane, as well as a maximum of 2 years to use it after its creation. The main use of the certificates is to report on compliance with the targets for reducing emissions and the market share of renewable energies in transport.

Any consumer can obtain biomethane through their supply contract; however, it is necessary to emphasise that France incentivises the use of biomethane in transport, given that suppliers only keep 100% of the profit resulting from the economic transaction of the certificate in the event that the sale is made for use in mobility and only 25% if it is intended for heating or gas.

France has committed to ambitious targets in terms of the production of biogas, established in the Energy Transition Law, which entail making 10% of the natural gas consumed renewable by 2030.

In accordance with the Distribution Grid for Natural Gas in France (GRDF) [9], the production of biogas in 2016 was 3.6 TWh/year [10] of which 215 GWh/year were recovered as biomethane through its biomethanation plants.

This development goes hand in hand with the great growth of biogas recovery plants, which have increased from a total of 4 in 2013 to 30 in 2017, which means that this country has positioned itself as one of the most promising future markets.

7.2 Italy

Italy has a support system based on the creation of guarantees of origin for the use of biomethane in mobility. The fuel distributors are the final recipients of the certificates, given that they have to use them to accredit compliance with the targets for renewable energies in transport [11].

These certificates are obtained at the time when the projects inject the biomethane into the grid. Italian biomethane producers and fuel distributors establish bilateral contracts for the purchase and sale of renewable gas. Thus, when the producers make deliveries of renewable gas, the fuel distributors receive gas and certificates. The economic incentive reflected in the certificate depends on the biomethane production technology.

Italy currently has seven biomethane plants with a production of 50GWh, in contrast to the two plants it had in 2013.

7.3 Germany

Germany is an atypical case given that it does not have a feed-in tariff for the injection of biomethane into the gas grid, so the economic benefit only occurs due to the transaction of the guarantee of origin certificate.

The production of biomethane has tax incentives that result in the discount of tolls for injecting into the natural gas distribution grid.

The biomethane producers sell natural gas on the wholesale market through bilateral contracts for the purchase and sale of gas. The producer receives a certificate for each kWh at the time of the injection of renewable gas and transfers it to the suppliers in accordance with the amounts of renewable gas contracted [12].

The supplier sells the consumer the natural gas along with the certificates; furthermore, any consumer, whether industrial, individual or collective, can request green gas through their supply contract.

Transactions on the market of guarantee of origin certificates are only possible between gas suppliers, but not for end consumers. However, any consumer can request green gas through their supply contract.

Germany is the European leader in terms of biomethane production plants, with 194 units that generate 9.4 TWh/year. Also, the country has 103 gas fuel stations that supply 100% biomethane [13].

7.4 The Netherlands

In the Netherlands there is a feed-in tariff for the injection of biomethane into the gas grid and guarantees of origin. On the other hand, all of the biomethane projects and some biogas projects have subsidies for investment costs through corporate tax as producers of renewable energy.

The renewable natural gas market in the Netherlands has one peculiarity that distinguishes it from the other European countries as the injection of each m3 of biomethane into the transport grid results in a tradable certificate, which can be sold among registered traders in the gas system, and it does not have to be linked to a physical flow of gas. For sales of biomethane outside of the gas grid, a non-tradable certificate is generated, whose recipient is the end consumer; it is not separable from the physical flow of gas.

The current biogas market in the Netherlands amounts to 1.1 TWh/year, although most of this biogas, thanks to different support systems such as feed-in premium, is recovered through cogeneration plants for use in electricity and heat. Only 15% of this production is recovered as biomethane for its injection into the grid and for obtaining guarantees of origin.

In total, there are 26 biomethane plants in the Netherlands which supply 91 gas fuel stations with 100% biomethane; they represent more than half of the country's supply infrastructure grid.

The Netherlands National Renewable Energy Action Plan [14] establishes the targets for the use of biomethane in mobility. It is expected that the road map will offer great potential to biomethane as a fuel. The main gas producers in the country are already offering the supply of renewable gas to users, although most of it is through the import of renewable gas certificates from the United Kingdom.

7.5 The United Kingdom

The injection of biomethane into the gas grid in the United Kingdom has a feed-in tariff, but it does not have specific incentives for mobility.

Producers sell biomethane by injecting it into the natural gas transport grid, or they sell it directly liquefied or compressed for transport. The producer receives the certificate at the time of the injection of biomethane into the grid, and they either transfer it to the suppliers in accordance with the contracts signed for the purchase of green gas or they transfer it without depending on physical gas flows.

The suppliers can sell certificates and gas to an end consumer or to other suppliers. Once the certificate reaches the end consumer, it cannot be exchanged again, and it is redeemed.

These transactions are possible with suppliers and/or consumers from outside the United Kingdom, always through interconnections and verifying that the physical gas flow took place between both countries.

The production of biomethane has increased significantly since 2013, reaching 3.75 TWh/year in 2017, which are produced in a total of 85 biomethanation plants.

In general, significant growth of the injection of biomethane is expected in order to achieve the targets set: an increase of 60% in the injection of biomethane into the grid, as well as an increase of the installed capacity to 5 TWh/year [15] due to new projects that are undergoing development.

7.6 Denmark

Denmark has a support system that is very favourable for the injection of biomethane into the gas grid through a feed-in premium scheme, which consists of three tariffs:

Tariff 1: Variable tariff. It began to be used in 2016, and its value drops by 1/5 each year, so it will be eliminated in 2020.

Tariff 2: Variable tariff. It is inversely proportional to the price of natural gas.

Tariff 3: Fixed tariff. Mainly, priority is given to injection into the grid and transport as the only viable decarbonisation option.

Additionally, in Denmark there is another feed-in premium scheme for the sale of biomethane for direct consumption in the transport or industrial sectors.

Denmark attained biomethane production of 898 GWh/year, with 50% injected into the grid. Although there is certain uncertainty over the future development of regulation after 2020, there is no reason to believe that trends are going to change as the Danish Energy Agency calculates that 5 TWh/year will be injected into the grid in 2022.

7.7 Sweden

Sweden has not implemented any certification system for biomethane and nor does it have feed-in tariff or feed-in premium systems [16]. Nevertheless, the total production of biomethane during 2016 was 1.3 TWh.

The national operator of the natural gas transport grid has implemented the Green Gas Principle, a virtual balancing point for renewable gas through bilateral agreements, which acts as an initial phase due to the lack of guarantees of origin.

The incentive scheme strongly supports biomethane for transport purposes through tax exemptions for vehicles, as well as exemptions from the CO_2 quota for biomethane as a fuel. All of this explains why Sweden is one of the most promising countries for the use of and harnessing of biomethane in mobility. An example of this is that almost all of its gas fuel stations supply 100% biomethane.

7.8 Comparative analysis

The European countries analysed have different levels of development of the renewable gas industry in terms of plants, infrastructures and technical requirements.

There are different formulas for carrying out the injection and connection to the grid, which differ in terms of the party that is required or responsible for the different phases of the injection and the existing obligations in order to prioritise, or not prioritise, biomethane over conventional natural gas.

Between 2013 and 2017, several support mechanisms for biomethane were implemented in most of the countries analysed, and consequently major growth is observed both in terms of the number of plants and the GWh produced between those years [17] (**Table 1**).

		France	Italy	Germany	Netherlands	United Kingdom	Denmark	Sweden
2013	Biomethane plants	4	2	154	23	6	3	54
	Biomethane production (Gwh)	20	15	7231	900	n/a	n/a	900
2017	Biomethane plants	30	7	194	26	85	22	63
	Biomethane production (Gwh)	215	60	9400	1100	3750	898	1297
	Number of CNG stations	87	1246	857	175	20	17	181
	No. of 100% Bio stations	6	1	103	91	2	14	141
	Obligatory quotas (%)	10% Achieved. NG End 2030	10% bio transport 2020	20% VNG (achieved)	100% bio transport 2020	n/a	No	30% Achieved NG end 2030
	Tax exemptions			Discounts on tolls in	Corporate tax		For use in heating	Exempt from taxes
	Subsidies, soft loans, investment grants and/or OPEX	Incentives depending on project						Investment incentives depending on project

Table 1.
Level of development of renewable gas industry (2013–2017).

				France	Italy	Germany	Netherlands	United Kingdom	Denmark	Sweden
Incentives for consumption	Injection incentives	BIOMETHANE	FiT/FiP for biomethane	✓	✓	✗	✓	✓	✓	✗
			Tax incentives (biomethane)	✗	✗	✓	✓	✗	✓	✓
			Extra incentive for use in mobility	✓	✓	✓	✗	✗	✓	✓
	Integrated model		Biomethane certificates	✓	✓	✓	✓	✓	✓	✗

Table 2.
International experience of support mechanisms.

DOI: http://dx.doi.org/10.5772/intechopen.91344

We have analysed the different systems from the point of view of cost recovery for biomethane producers. While in France producers only receive FiTs for the sale of biomethane, in Germany they only receive the profit generated by the certificates, and in Denmark, Italy, the Netherlands and the United Kingdom, they receive FiTs plus the profit generated through the sale of certificates (**Table 2**).

8. Conclusions

The sustainable transport of the future depends on incorporating decarbonised gases that make it possible to achieve the demanding climate targets.

Biomethane, synthetic gas or power-to-gas and hydrogen are the technologies that we can avail of in order to decarbonise conventional gas, but for that purpose, we need to overcome a significant number of drawbacks such as the reluctance of stakeholders in the current market to invest in decarbonised gas and a regulatory framework that has not been designed with this new reality in mind.

However, not everything is an obstacle; the use of natural gas as fuel is a mature technology for all kinds of vehicles that is gaining real momentum, especially for heavy-duty transport and maritime transport. These sectors do not have an environmentally friendly and efficient alternative for replacing conventional fuels. The only technologically mature option at present is natural gas and the progressive incorporation of renewable gas, biomethane, hydrogen or synthetic gas.

Additionally, the use of biomethane does not entail any risk to the system and nor does it require making any change to the vehicles or users, and it is an option that will make it possible to fulfil the obligations arising from the new Directive on renewable energies and reduce the foreseeable impact of the sanctions that will be faced by vehicle manufactures that do not manage to meet the targets set by Europe.

This chapter has offered an analysis of the incentive schemes of seven European countries that make it possible to overcome the aforementioned obstacles in order to make biomethane a reality, and although we have seen that there are different formulas we can see that there are numerous common policies.

It seems internationally accepted that the essential step towards introducing biomethane in transport is the definition of a system of certificates or guarantees of origin for the injection into the grid that is similar to the existing one for renewable electricity.

We also see that most of the countries analysed accompany the certificates with subsidies for biomethane producers and with tax benefits.

However, these subsidies could be reduced by providing the certificates with sufficient value in order to overcome the resistance of investors.

In a context where vehicle manufacturers face paying fines worth millions, the certificate could acquire an important value provided that it could be used to accredit compliance with the emission levels.

To do so, the methodology that the European Union bases its regulation on would have to be updated with a well-to-wheel approach, which takes into consideration the origin of the fuel and not only exhaust emissions.

Author details

Eugenia Sillero Maté, Carla García, Paloma Martínez Ramos
and Elena Martínez Calvo*
GASNAM, Spain

*Address all correspondence to: e.martinez@gasnam.es

References

[1] Boltz W, Jones C. The Decarbonisation of the European Energy Market. Gas, Hydrogen and an Affordable and Competitive European Energy Market. Naturgy Foundation; 2019

[2] Muñoz F. 2021 CO_2 Targets Would Generate €34 Billion Euros in Penalty Payments within Europe—JATO. 2019. Available from: https://www.jato.com/2021-co2-targets-would-generate-e34-billion-euros-in-penalty-payments-within-europe/ [Accessed: 18 October 2019]

[3] Foundation for the Promotion of Industrial Innovation. Higher School of Industrial Engineers of the Polytechnic University of Madrid. 2019. Report on Vehicle Emissions that Affect Air Quality. Available from: https://gasnam.es/wp-content/uploads/2019/09/INFORME-DE-EMISIONES-DE-VEH%C3%8DCULOS-QUE-AFECTAN-A-LA-CALIDAD-DEL-AIRE_GASNAM.pdf

[4] Thinkstep. Greenhouse Gas Intensity from Natural gas in Transport. 2017. Recuperado de http://ngvemissionsstudy.eu/

[5] Bouter A, Melgar J, Telner C. Etude ACV de véhicules roulant au GNV et bioGNV. 2019. Available from: https://www.ifpenergiesnouvelles.fr/article/analyse-du-cycle-vie-acv-des-vehicules-fonctionnant-au-gnv-et-biognv

[6] NGVA Europe & EBA European Biogas Association. g-mobility: Driving Circular Economy in Transport. 2018. Available from: https://www.ngva.eu/wp-content/uploads/2019/07/circular-economy-leaflet_190718.pdf

[7] Khan MI. Identifying and addressing barriers for the sustainable development of natural gas as automotive fuel. 2017

[8] GASNAM, SEDIGAS, AEBIG. Regulatory Benchmarking of Renewable Gas in Europe. Madrid: KPMG; 2019

[9] GRDF. Gestion des garanties d'origine biomèthane- Rapport anuel d'activité. 2018

[10] GRDF. Panorama du Gaz renouvelable en 2018. 2019

[11] GSE. Energie rinnovabili al 2020 Scenari tendenzial. 2018

[12] FNR. Bioenergy Prom: Biomethane production in Germany. 2017

[13] NGVA Europe Gas Stations Map. 2019. Available from: https://www.ngva.eu/stations-map/

[14] Vertogas BV. Pieces of the Certificate Puzzle. 2017

[15] JIN Climate and Sustainability, Groningen. Green Gas Certification Scheme: A Level Playing Field for the European Biogas and Biomethane Markets. Bremen: Carl von Ossietzky University, Oldenburg; Jacobs University; 2015

[16] Swedish Gas Association. Proposal for Natural Biogas Strategy 2018. 2018

[17] EBA. Statistical Report of the European Biogas Association (EBA). 2017

Chapter 8

Sustainability Analyses for Hydrogen Fuel Cell Electric Vehicles

Hüseyin Turan Arat, Bahattin Tanç and Nevzat Özaslan

Abstract

One of the most important energy sources for well and availability is undoubtedly hydrogen. Both the production capacity and the usability of hydrogen in various industrial sections are continuing to increase significantly and sustainable. With this high trend, it is unthinkable that hydrogen is not a part of the automotive industry. The aim of this book chapter is giving info on the hydrogen fuel cell electric vehicles potential which have developing e-motor, charging capabilities and battery capacities; and illustrate the upgrade their stars on industry. Basic fundamentals, structural features, merits & demerits and energy efficiency analyzes will be done. In addition, for the future of sustainable mobility, the future milestones will be discussed and the current economical manner will be discussed in terms of life cycle assessment and environmental perspective. As a result, a comprehensible hydrogen fuel cell electric vehicle potential will be examined in both the automotive industry and all stakeholders. FCEVs on availability of autonomous vehicles will also be discussed, too.

Keywords: sustainability analyses, fuel cell electric vehicle, life cycle assessment, energy analyses

1. Introduction

Humanity has tended to supply energy from fossil fuels since the industrial revolution. Although this trend has paved the way for great advances in the name of humanity and has provided the necessary energy needs, new searches have been taken due to the limited availability of fossil fuels and the regular increase in human energy needs. As a result of these efforts, alternative energy has become the most important resource in terms of sustainability by being more environmentally friendly and renewable than fossil fuels. Considering the harm that fossil fuels cause to the environment due to carbon emissions, electric vehicles and hydrogen fuel cells have also come to the fore as environmental sustainability with zero emission values. The OECD announced that the health expenditure resulting from air pollution caused by vehicles was approximately 550 T euro [1]. For 2018 alone, total carbon emissions were 954,677 million tons [2]. The carbon emission, which was previously at the level of billion tons, has shown a downward trend in recent years with the introduction of alternative energy sources. In this regard, states have ended the policy of obtaining fossil fuel from alternative energy sources which are completely environmentalist.

The European Union has adopted the policy of reducing GHG by 40% by 2030 and by 80% by 2050 [1]. As a result of this policy, the European Union states have decided to support the infrastructure of electric vehicles and hydrogen fuel cells in order to reduce carbon emissions in their own countries. Germany, one of the most important examples of this, has commissioned 34 hydrogen refueling stations so far and aims to increase this number to 400 by 2023 [3]. According to the research conducted by Mckinsey company, it is stated that an investment cost of approximately 3 Billion € is required for 1 M hydrogen fuel cell vehicle [3]. Asian states have also added to their policies the support of hydrogen fuel cell vehicles to solve the GHG problem. The Japanese government, 320 hydrogen filling plant to put into operation by 2025 and 800,000 vehicles on the road aims to remove by to 2030.

Due to the goal of reducing carbon emissions, the most important alternative fuel is the hydrogen which obtained from renewable energy sources. The idea of using hydrogen as a fuel is not a new idea but new technologies have enabled hydrogen to be sustainable. In order to use hydrogen, which is present as compounds in nature, it needs to be purified. The energy sources of the methods used to purify hydrogen should also be examined. Because if these methods use an energy source that leaves a carbon footprint, the hydrogen produced by this method cannot be considered as an alternative energy source. There are two types of hydrogen purification processes that are actively used. These are obtained from methane in natural gas and electrolysis, a method of obtaining from water [4].

Nowadays, hydrogen, which will be used as a clean alternative fuel, should be purified by electrolysis. The electricity used in the electrolysis method should also be provided from renewable energy sources. Otherwise, the GHG emission will be higher than for internal combustion engines [4]. The technologies of hydrogen fuel cell vehicles have not reached maturity yet, but electric vehicles, the most powerful alternative, have reached maturity. Even though electric vehicles have reached maturity, there are a few issues where hydrogen fuel cell vehicles are still superior, and they still make hydrogen-fuel cell vehicles the best alternative. The major advantage of hydrogen fuel cell vehicles over electric vehicles is that they offer a range as long as internal combustion engines. Hydrogen fuel cell vehicles also have the advantage of filling as soon as internal combustion engines. In order to increase the market share of hydrogen fuel cell vehicles, manufacturers expect the demand to be created and the infrastructure to be strengthened, while consumers want the vehicles to be produced, parts supply shortage and infrastructure to be improved. States, on the other hand, have accelerated the infrastructure studies to break this paradox. In this way, producers will start production as infrastructure is waiting without demand, and consumers will start to buy cars through government investments.

Even though hydrogen refueling stations are the first investment that comes to mind, and can listed the most important investments of state and private companies as water, using renewable energy (wind, solar, wave energy) by electrolysis method and laying pipes for the transportation of hydrogen. Other advantages of hydrogen fuel cell vehicles compared to vehicles using an internal combustion engine are that they are vibration free, do not require shifting and are quiet, while the cons can be considered to waive the luggage volume for the tanks to be loaded with hydrogen [5]. The current production costs of hydrogen fuel cell vehicles are still higher, as compared to electric vehicles, because their technology does not reach maturity [5]. The Toyota Mirai reached a 500 km range with a 1.6 kW engine and a high pressure 5 kg hydrogen tank [6]. The current price of hydrogen ranges from $ 12.85 to $ 16. On average, the price of hydrogen per kg is $ 13.99 [7]. This price is expected to drop to at least $ 4 per kilo with the maturation of electrolysis technology [8]. Hydrogen gas is not more dangerous than other fuels. Naturally, all fuels must be flammable and hydrogen is a flammable fuel too. Hydrogen was also used as city gas before natural gas came to USA [9]. One of the less dangerous of hydrogen can be expressed with its less density. In this way, in case of any leakage, it will increase in the atmosphere since the hydrogen density is low [9].

In spite of this critical info about hydrogen, fuel cells and vehicles; a foresight future of hydrogen and related items has been published as a report [7] by International Energy Agency, right now. A critical future perspective is observed very detailed on hydrogen for G20 meeting. All fields of hydrogen were summarized and enlighten descriptions were given for future aspects and sustainability of hydrogen in terms of industrial applications and transportation sector.

In this book chapter, authors aim to given the fundamentals of fuel cell vehicles in addition with sustainably manner. Also, an example of fuel cell vehicle energy performance was illustrated with AVL Cruise, too. And lastly important future recommendations were mentioned.

2. Fuel cells and vehicle applications

Energy is the wanted key component in all industrial applications for everyone. In case of sustainability, one of the most important necessities is energy source. In terms of source, hydrogen is the one of the most abundant energy carrier in the world. Addition of this importance, fully environmental friendly structure of hydrogen that it is becoming with its nature, prefers as a carbon free source, too.

Basically, fuel cell can be expressed as; electrochemical device which converts the chemical energy to electric energy by using different fuels. There are six well known fuel cells and named as Polymer Electrolyte Membrane (PEM), Alkaline (AFC), Phosphoric Acid (PAFC), Solid Oxide (SOFC) and Molten Carbonate (MCFC). The main differences of each other are operating temperatures and electrodes. In authors previous study [10] a detailed illustration of fuel cells was shown and refigured in **Figure 1**.

In transportation sector, generally PEM type fuel cell is preferred. The main reason of this selection is depended on the operation temperature. Usually, fuel cell vehicles driven with PEM cells. Various automotive manufactures on fuel cell

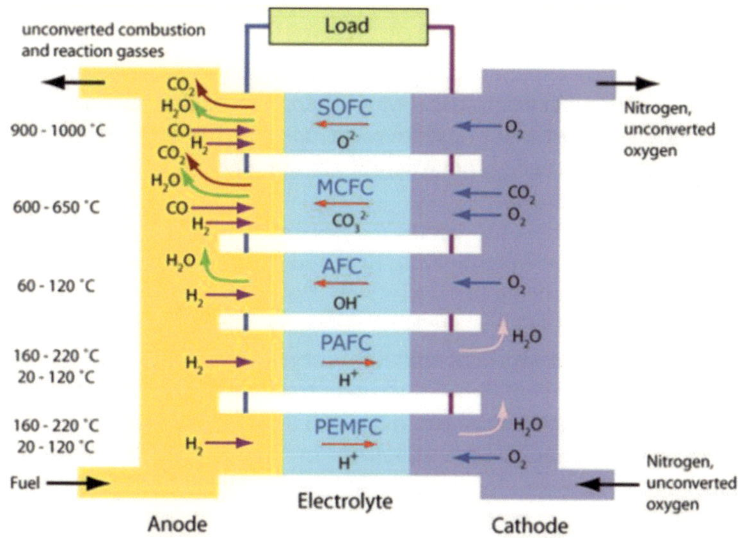

Figure 1.
Fuel cells structures and operation temperatures [10, 11].

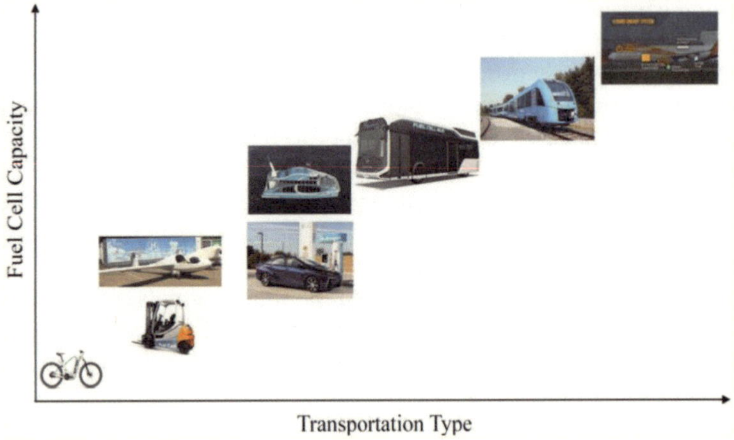

Figure 2.
Fuel cells in transportation industry.

vehicle (Toyota, Hyundai, Honda, and Mercedes [on production not concepts]) tended to use PEM with 45–60% efficiencies.

On the other hand, not only automotive sector but also all fields of transportation industry, given importance on fuel cells. **Figure 2** illustrated the sample products of transportation industries which includes land, marine and air vehicles.

In addition to all these, especially in the technological development process, the studies carried out in fuel cells have increased importance in recent years. Especially in Europe, important moves are made on fuel cell propulsion in marine and train transport and its infrastructure is created for them. On the other hand, large aviation companies are working on driving aviation with electricity and fuel cell. At this point, PEM fuel cells appear to be the most suitable way in terms of operating parameters. Since such fuel cells use hydrogen as fuel; all governments interested in the subject that are preparing and investing in hydrogen refueling stations. As such, fuel cell related sectors will emerge as rising stars in the next 20 years.

3. Sustainability analyses

Sustainability is one of the most important analyzes for each manufactured production and developed systematic. For all theories, the definition and analysis of sustainability consists of three basic perceptions: Environmental, Economic and Social factors. Hydrogen energy sustainability assessments generally give positive results in all three perceptions.

In environmental manner; life-cycle analysis, material flows, resource accounting and ecological footprint headings can be listed and should analyzed. For hydrogen powered fuel cell vehicles, Life Cycle Assessment (LCA) analyses were detailed analyzed by [12–14].

In terms of economic analyses of sustainability; cost/benefit analysis, modeling, regressions and scenarios were done by various researchers for fuel cell vehicles. General opinion of these studies summarized as, the costs of fuel cell vehicles with indeed materials were expensive nowadays but in the short run, combined with newly produced materials and increased efficiency, hydrogen supply–demand balance in transportation sector would change positively, prices will decrease and demand will increase.

When it comes to social effects, it is seen that, social effect is in a much better state than other effects. In particular, the fact that mass production FCEVs take place in the transportation sector and improves sales figures multiply every year; the social perspective of people perceived positively. In the surveys, the majority of the subjects stated that they could buy fuel cell vehicles in the near future. In addition, governments, especially Japan, have started to organize their own programs on the usage of hydrogen energy and set up departments on energy ministries. The G-20 summit (2019) is an important touchstone for this issue.

In addition, almost all well-known car manufacturers produce their concept fuel cell vehicles. In this case, it significantly strengthens social sustainability. This book chapters references were prepared for readers to gaining more information of this critical issue [1–50].

Considering environmental, technological and social factors; the sustainability of fuel cell electric vehicles will gradually increase its importance over the next 20–30 years. Parameters such as increasing population, decreasing energy sources, carbon foot-print minimization, consumer demand, the evolution of technology into electrical propulsion systems, etc. are clues that the sustainability of FCEVs will be continue.

4. A simulational example of fuel cell vehicle performance analysis

In this section, an example of the simulation stages and results of fuel cell electric vehicle modeled with AVL-Cruise simulation program were given. The AVL (Anstalt für Verbrennungskraftmaschinen List) company, founded in 1948 by Professor Hans LIST in Austria, has become one of the world's leading engineering simulation, measurement, application, modeling and realization companies [15]. One of the main simulation tools is AVL-Cruise. With this program the performance, emission and energy distribution results of any conventional, hybrid, fuel cell and electric vehicle can be concluded with different simulation and code software integrations (Matlab/Simulink, C ++, etc.).

Figure 3 shows the simulation interface of an exemplary fuel cell electric vehicle in AVL Cruise. From the modules bank, it can be chosen from a wide variable options, including all mechanical devices and interactive program options, fuel cell, electric motors, main power units, power train and gearbox, driving cycles, cockpit and driver details.

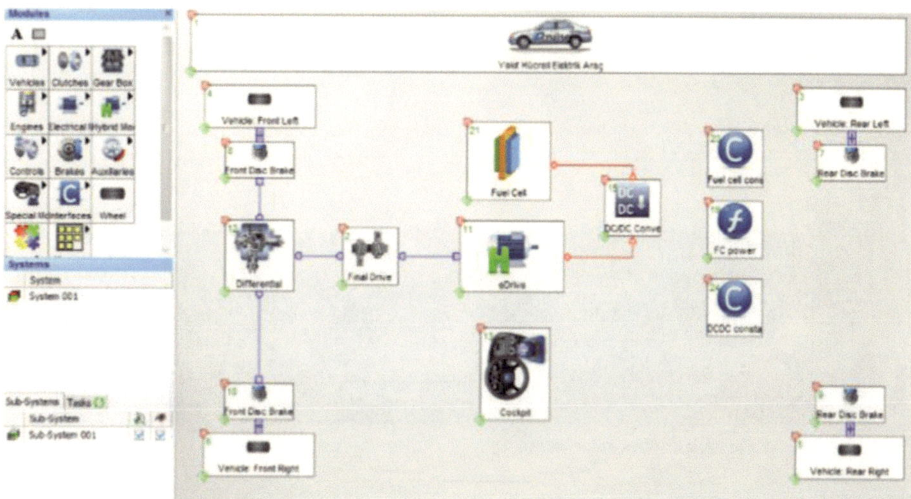

Figure 3.
Example of AVL schematic [15].

The components selected from the module pool must be inter-connected each other. In **Figure 3**, the connection system was shown and colors indicated as blue: mechanical, red: electronic colors should be connected as shown in **Figure 4**.

All simulation programs have specific operating systems. In the system where codes and certain mathematical calculations [15] will be performed, the input conditions must be determined and entered as data. In order to give accurate approximate results for the program, some of the required vehicle specifications entered the program must have been previously measured (tested) or either measured data from the manufacturer (factory data). After obtaining the accuracy report in the system, it can proceed to the simulation execution section shown in **Figure 5**. An important point is noted here, that the vehicle, the driving cycle and the cockpit and the driver's part must be specified in the project sub-stage.

When the results section is started, the main reporting results and the input data given according to different parameters and the simulated results are displayed in the tree image. For example, **Figure 6** shows instant energy input/output diagram (Sankey). As shown in the sections indicated by red letters in the figure; (a) result tree

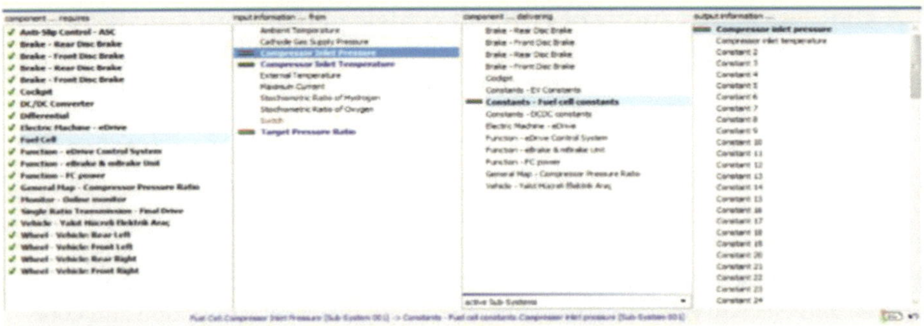

Figure 4.
Connection and communication units page example [15].

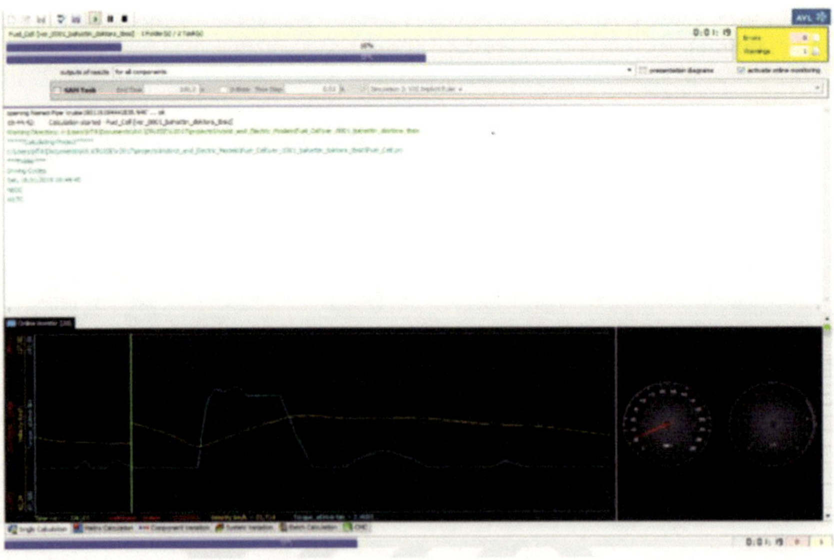

Figure 5.
Single computational analysis screenshot [15].

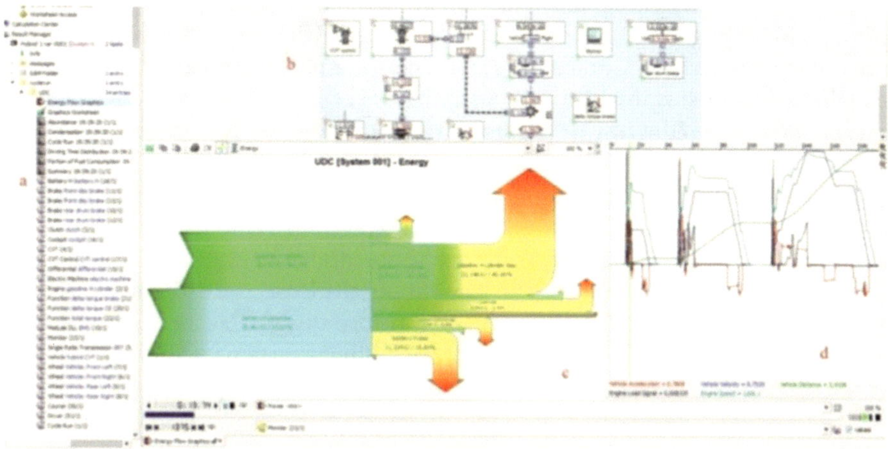

Figure 6.
Example model Sankey diagram result example of the vehicle [15].

chart; (b) instant power exchange on the vehicle; (c) Sankey diagram; and (d) the chart in where the instant results can be seen according to the selected driving cycle.

5. Conclusion

In this book chapter, authors tented to give the last critical info's about; the fundamentals of the development of hydrogen energy, the necessity of producing with renewable energy instead of fossil fuels, the importance of reducing the carbon foot print for emissions, the reasons for governments to consider hydrogen as an alternative fuel, the required steps had been taken, and the investments should be taken into consideration.

The usage of hydrogen in fuel cell vehicles and their applications on various transportation sectors were explained. The sustainability of hydrogen fuel cell vehicles mentioned detailed with newest literature and reported studies.

An essential example of energy distribution and efficiency analyses were consisted for a modeled hydrogen fuel cell vehicle which were made by using AVL Cruise program.

In the future perspective, either with mentioned references of this book chapter is expected that, the potential of hydrogen in the transport sector will increase its potential and the cost of materials would decrease. In this way, it is envisaged that the range will be comparable with internal combustion engines and carbon emissions with electric vehicles.

As an important result of fuel cell vehicles phenomenon; it will play an important role in autonomous vehicle sector, too.

6. Future recommendations

Fuel cell electric vehicles will be a crucial milestone in the future in line with the sustainability of the Earth's environmental ecosystem and therefore the decisions of many countries' governments. Developments in the economic sustainability of fuel cell vehicles are also expected to rise in parallel with this situation.

Reducing costs in hydrogen production methods, establishing a large network of Hydrogen Refueling Station and increasing fuel cell efficiency are among the main objectives of researchers. Considering that energy, environment and economy are the main factors of sustainability, diversity in fuel cell membranes, reduction of emission gases and reduction of preliminary costs are possible in the next quarter century.

The main materials such as membranes, plates and electrolytes will be the most appropriate study subjects for the future fuel cell systems. And then, to ensure sustainability aspects; the efficiency, cost, environmental effects and social manners will follows the materials studies. The last but not the least, the economy of hydrogen enlightened the visionary perspective for future sustainable alternative energy. For this reason, the authors strongly recommend that everyone work selflessly to give the necessary importance for hydrogen in all over the world.

Acknowledgements

This study was conducted with AVL CRUISE. Authors acknowledge to AVL-AST, Graz, Austria to provide these simulation tools under university partnership program.

Conflict of interest

The authors declare no conflict of interest.

Nomenclature

AVL	Anstalt für Verbrennungskraftmaschinen List
AFC	Alkaline Fuel Cell
FCEV	Fuel Cell Electric Vehicle
GHG	Greenhouse Gases
IEA	International Energy Agency
LCA	Life Cycle Assessments
MCFC	Molten Carbonate Fuel cell
OECD	Organization for Economic Co-operation and Development
PAFC	Phosphoric Acid Fuel Cell
PEM	Polymer Electrolyte Membrane (Fuel Cell)
SOFC	Solid Oxide Fuel Cell

Author details

Hüseyin Turan Arat[1*], Bahattin Tanç[2] and Nevzat Özaslan[2]

1 Department of Mechatronics Engineering, Faculty of Engineering and Natural Sciences, İskenderun Technical University, İskenderun, Hatay, Turkey

2 Department of Mechanical Engineering, Faculty of Engineering and Natural Sciences, İskenderun Technical University, İskenderun, Hatay, Turkey

*Address all correspondence to: hturan.arat@iste.edu.tr

References

[1] Borén S, NY H. A strategic sustainability analysis of electric vehicles in EU today and towards 2050. International Journal of Environmental and Ecological Engineering. 2016;**10**(3):294-302

[2] Carbon footprint CO_2 emissions; Full Scope. April 2019. Available from: https://www.sanofi.com/-/media/Project/One-Sanofi-Web/Websites/Global/Sanofi-COM/Home/common/docs/download-center/Carbon-footprint-2019.pdf?la=en&hash=87BCE572F5CA7FB2F617774240DA4614E084F0DB [Accessed: 18 December 2019]

[3] Brandon NP, Kurban Z. Clean energy and the hydrogen economy. Philosophical Transactions of the Royal Society A: Mathematical, Physical and Engineering Sciences. 2017;**375**(2098):20160400

[4] Rosenfeld DC, Lindorfer J, Fazeni-Fraisl K. Comparison of advanced fuels—Which technology can win from the life cycle perspective? Journal of Cleaner Production. 2019;**117879**:238

[5] Staffell I et al. The role of hydrogen and fuel cells in the global energy system. Energy & Environmental Science. 2019;**12**(2):463-491

[6] Purnima P, Jayanti S. Fuel processor-battery-fuel cell hybrid drivetrain for extended range operation of passenger vehicles. International Journal of Hydrogen Energy. 2019;**44**(29):15494-15510

[7] IEA. The future of hydrogen. The IEA technical report for G20 meeting; 2019

[8] Nagasawa K et al. Impacts of renewable hydrogen production from wind energy in electricity markets on potential hydrogen demand for light-duty vehicles. Applied Energy. 2019;**235**:1001-1016

[9] Dincer I. Hydrogen and fuel cell technologies for sustainable future. Jordan Journal of Mechanical and Industrial Engineering. 2008;**2**:1

[10] Tanç B et al. Overview of the next quarter century vision of hydrogen fuel cell electric vehicles. International Journal of Hydrogen Energy. 2019;**44**(20):10120-10128

[11] Cambridge University. Department of online resources for the teaching and learning of materials science. DoITPoMS; 2018. Available from: https://www.doitpoms.ac.uk/tlplib/batteries/battery_characteristics.php [Accessed: 18 December 2019]

[12] Veziroğlu A. Hydrogen Powered Transportation. Xlibris; February 2017. ISBN: 1524582948

[13] Dincer I. Environmental and sustainability aspects of hydrogen and fuel cell systems. International Journal of Energy Research. 2007;**31**(1):29-55

[14] Granovskii M, Dincer I, Rosen MA. Life cycle assessment of hydrogen fuel cell and gasoline vehicles. International Journal of Hydrogen Energy. 2006;**31**(3):337-352

[15] AVL User guide. AVL simulation tools; 2018. Available from: www.avl.com

[16] Creti A, Kotelnikova A, Meunier G, Ponssard JP. A cost benefit analysis of fuel cell electric vehicles; 2015. [Research Report] ffhal-01116997f

[17] Stephens TS, Birky A, Gohlke D. Vehicle Technologies and Fuel Cell Technologies Office Research and Development Programs: Prospective Benefits Assessment Report for Fiscal Year 2018. Argonne, IL (United States): Argonne National Lab. (ANL); 2017

[18] Shell, Energy of the Future? Technical Report; 2018. Available

from: https://hydrogeneurope.eu/sites/default/files/shell-h2-study-new.pdf [Accessed: 18 December 2019]

[19] Nieuwenhuis P, Wells P. New business models for alternative fuel and alternative powertrain vehicles; an infrastructure perspective. In: New Business Models for Alternative Fuel and Powertrain Vehicles. 2012. Available from: https://www.oecd.org/futures/New%20Business%20Models%20for%20Alternative%20Fuel%20and%20Alternative%20Powertrain%20vehicles.pdf [Accessed: 18 December 2019]

[20] Eaves S, Eaves J. A cost comparison of fuel-cell and battery electric vehicles. Journal of Power Sources. 2004;**130**(1-2):208-212

[21] The International Consortium for Fire Safety. Health and the environment. In: Safety Issues Regarding Fuel Cell Vehicles and Hydrogen Fueled Vehicles. 2003. Available from: https://dps.mn.gov/divisions/sfm/programs-services/Documents/Responder%20Safety/Alternative%20Fuels/FuelCellHydrogenFuelVehicleSafety.pdf [Accessed: December 2019]

[22] Brouwer J. On the role of fuel cells and hydrogen in a more sustainable and renewable energy future. Current Applied Physics. 2010;**10**(2):S9-S17

[23] Alfonsín V et al. Simulation of a hydrogen hybrid battery-fuel cell vehicle. Dynamis. 2015;**82**(194):9-14

[24] The Society of Motor Manufacturers and Traders, Hydrogen Fuel Cell Electric Vehicles; 2019. Available from: https://www.smmt.co.uk/wp-content/uploads/sites/2/2019.03.11-SMMT-FCEV-guide-FINAL.pdf [Accessed: 18 December 2019]

[25] Australian Government, National Hydrogen Strategy; 2019. Available from: https://www.industry.gov.au/data-and-publications/australias-national-hydrogen-strategy [Accessed: 18 December 2019]

[26] Herb F et al. Theoretic analysis of energy management strategies for fuel cell electric vehicle with respect to fuel cell and battery aging. In: World Electric Vehicle Symposium and Exhibition (EVS27). IEEE; 2013. pp. 1-9. Available from: https://ieeexplore.ieee.org/document/6915049 [Accessed: 18 December 2019]

[27] Hwang J-J. Sustainability study of hydrogen path ways for fuel cell vehicle applications. Renewable and Sustainable Energy Reviews. 2013;**19**:220-229

[28] Guirong Z, Houyu L, Fei H. Propulsion control of fuel cell electric vehicle. Procedia Environmental Sciences. 2011;**10**:439-443

[29] Berkeley University, Western Washington State Clean Cities Webinar; 2017. Available from: https://www.pscleanair.gov/DocumentCenter/View/3032/Washington-State-Briefing-2017?bidId= [Accessed: 18 December 2019]

[30] Helm D. The European frame work for energy and climate policies. Energy Policy. 2014;**64**:29-35

[31] Offer GJ et al. Comparative analysis of battery electric, hydrogen fuel cell and hybrid vehicles in a future sustainable road transport system. Energy Policy. 2010;**38**(1):24-29

[32] European Commission, Energy road-map 2050; 2011. Available from: https://ec.europa.eu/energy/sites/ener/files/documents/roadmap2050_ia_20120430_en_0.pdf [Accessed: December 2019]

[33] Wanitschke A, Hoffmann S. Are battery electric vehicles the future? An uncertainty comparison with hydrogen and combustion engines. Environmental Innovation and

Societal Transitions. 2019. https://doi.org/10.1016/j.eist.2019.03.003

[34] Emonts B et al. Flexible sector coupling with hydrogen: A climate-friendly fuel supply for road transport. International Journal of Hydrogen Energy. 2019;**44**(26):12918-12930

[35] Kurtz JM et al. Fuel Cell Electric Vehicle Durability and Fuel Cell Performance. Golden, CO (United States): National Renewable Energy Lab. (NREL); 2019

[36] Murugan A et al. Measurement challenges for hydrogen vehicles. International Journal of Hydrogen Energy. 2019;**44**(35):19326-19333

[37] Chen Y, Melaina M. Model-basedt echno-economic evaluation of fuel cell vehicles considering technology uncertainties. Transportation Research Part D: Transport and Environment. 2019;**74**:234-244

[38] Wang S et al. Prioritizing among the end uses of excess renewable energy for cost-effective green house gas emission reductions. Applied Energy. 2019;**235**:284-298

[39] Matute G, Yusta JM, Correas LC. Techno-economic modeling of water electrolyzers in the range of several MW to provide grid services while generating hydrogen for different applications: A case study in Spain applied to mobility with FCEVs. International Journal of Hydrogen Energy. 2019;**44**(33):17431-17442

[40] Michalski J, Poltrum M, Bünger U. The role of renewable fuel supply in the transport sector in a future decarbonized energy system. International Journal of Hydrogen Energy. 2019;**44**(25):12554-12565

[41] Tanç B. Hidrojen Yakit Hücreli Hibrit Elektrikli Araç İçin Destek Bataryasinin Enerji Dağilimi Ve Araç Performansi Üzerindeki Etkilerinin Analizi [PhD thesis]. Iskenderun Technical University; 2019

[42] Wang Y et al. Materials, technological status, and fundamentals of PEM fuel cells–a review. Materials Today. 2019. https://doi.org/10.1016/j.mattod.2019.06.005

[43] Haile SM. Materials for fuel cells. Materials Today. 2003;**6**(3):24-29

[44] Boureima F-S et al. An environmental analysis of FCEV and H2-ICE vehicles using the eco score methodology. World Electric Vehicle Journal. 2009;**3**(3):635-646

[45] Patterson J. Low Carbon Vehicle Partnership, Understanding The Life Cycle GHG Emissions for Different Vehicle Types and Powertrain Technologies; 2018

[46] Fathabadi H. Combining a proton exchange membrane fuel cell (PEMFC) stack with a Li-ion battery to supply the power needs of a hybrid electric vehicle. Renewable Energy. 2019;**130**:714-724

[47] Wang H, Gaillard A, Hissel D. A review of DC/DC converter-based electrochemical impedance spectroscopy for fuel cell electric vehicles. Renewable Energy. 2019;**141**:124-138

[48] Hart D. Hydrogen–a truly sustainable transport fuel? Frontiers in Ecology and the Environment. 2003;**1**(3):138-145

[49] Bruggink JJC et al. The economic feasibility of a sustainable hydrogen economy. In: Stelten D, Grube T, editors. Proceedings of the 18th World Hydrogen Energy Conference. 2010

[50] Javier DLCS, Cano U. Fuel Cell as Range Extender in Battery Electric Vehicles for Supply Chain Fleets. IntechOpen; 2016. DOI: 10.5772/62792

Section 4

Electricity

Chapter 9

Management of Innovation Performance on the Example of the Automotive Supply Chains

Ewa Stawiarska

Abstract

The chapter presents original research showing the relationship between expenditures on developing systems (including IT) supporting innovation management of supply chains of three international automotive companies and their innovation performance. A novelty in comparison to previous (cited) studies is that the calculated correlation coefficients show the significance of the link between the expenditures on components of the suppliers' innovation management system, that is, expenditure on the improvement of innovative competences of suppliers, and the expenditure on ICT, and innovation performance of automotive companies. By establishing data interdependence, elements of the management system that contributed the most to improving the innovation performance of three international auto motive companies over the past years were selected. Analyses of data may facilitate the management of expenditure on: internal R&D activities and support for external innovators. In this chapter, the author used the results of her previous studies. To collect additional data, the following methods were used: review of documents and diagnostic survey (technique: survey, tool: questionnaire). To analyse the data obtained, a statistical method and computer simulation methods were used. The research was based on quantitative secondary data[1] and estimates (given by respondents from the automotive companies as a percentage of a specific value contained in official documents). The data for the analysis came from publications and surveys.

Keywords: R&D, supply chain management, sustainability

1. Introduction

The supply chain will be understood as a group of companies that are suppliers and customers of each other. There is a flow of goods, information, knowledge, resources, funds, and innovative solutions between companies. The company that takes on the challenge of managing the supply chain is referred to as a leader or an integrator.

The innovation performance of companies that cooperate in supply chains may be increased by cooperation in three different dimensions: new technologies, new

---

[1] http://iri.jrc.ec.europa.eu/scoreboard.html

knowledge and skills, and new forms of cooperation [1]. To develop innovations in the open model (OM), the company that manages the supply chain (the leader) should incur expenditure on the following:

- R&D activities carried out together with suppliers

- Suppliers' innovation development programmes

- ICT implemented in order to streamline innovation development processes with suppliers

Making decisions on the amount of expenditure should be justified by statements on the sustainable development of supply chains and backed up by research. This paper presents the research (including research tools) and the results that facilitate decision-making.

The article confirms the thesis that the greatest automotive innovators develop new products with the use of the open innovation model (OI) and manage the innovation performance of their supply chains. Automotive companies are in need of the implementation of tools that support innovation productivity management (e.g., in order to optimise expenditure on the development of the open innovation model).

The basic objective of the article was to verify the usefulness of the research tool empirically, which-following adjustments-could be used in the practice of the innovation performance management of automotive companies and their supply chains.

The subsequent chapters of the paper analysed the development of measures of innovation performance and their application. There was a gap in studies concerning the measurement of innovation performance of cooperative companies. The studies that showed the importance of managing relations and resources of cooperative companies in the increase in innovation performance of supply chains were cited. Another research gap was identified. The researchers did not measure expenditure on the management of relations and resources of cooperative companies or expenditure on ICT tools that support such management in terms of their own innovation performance.

The article contains original research showing that the increase in expenditure on innovation management of suppliers results in an increase in the innovation performance of the leader in the supply chain. The analyses confirming this relationship are based on data on automotive companies, which have been at the forefront of the most innovative company rankings for years. The obtained results are new. Previous researchers showed that suppliers' management contributes to the increase in innovation performance. The presented research confirms this on the basis of data from three international companies, originating from different cultures, and thus differently managing relations with suppliers. It shows that even the surveyed companies (the world's greatest innovators) are struggling to implement a sustainable suppliers' development policy. The author believes that the results of this research will facilitate making investment decisions in the R&D activity and will lead to the transition to an open model of developing innovations together with suppliers. The author calls for the collection of data and the use of ICT for the analysis of expenditure on research and development. She proves that currently the data are scattered and inconsistent, which are, therefore, difficult to process. Finally, she postulates the preparation of a systemic model for the supply chain innovation management.

2. Measurement of innovation performance of enterprises

Innovation management can be defined as systematic planning, organising, implementing, and controlling activities carried out to develop and introduce new products and related processes [2]. In the literature, innovation management comes down to a management process consisting of the following three stages:

- Ideation, which refers to the very process of generating ideas and includes recognition of opportunities, analysis, generation of ideas, evaluation, and selection, as well as product concept creation [3].

- Development, the stage of new product development (NPD) or new service development (NSD), including calculating, modelling, and prototyping. A typical NPD process is presented in the Cooper Stage-Gate [4] model.

- Launch, at this stage the company is involved in the production and distribution, marketing, diffusion (e.g., licensing), learning, and getting experience from the project [5].

The ideation stage has a significant impact on the entire process of innovation [6]. The main result of this stage is the concept of a new product, which can then enter the formal process of technological development.

Difficulties in the management of ideation processes result from the scarcity of developed supporting instruments. Despite the rapid development of information and communication technologies (in particular search engines and computer databases), which significantly improves the process of searching for information, the actual emergence of an idea is still considered mainly an uncontrolled ad-hoc process [7]. In contrast, in the final stages, the engineers have many supporting tools for prototyping, such as CAD/CAE systems.

Idea generation sessions tend to be performed in groups rather than by individuals. Therefore, it is possible to identify a clear need to develop tools to facilitate the process of creating ideas. Innovation management tools are of interest to both companies and researchers. The objectives of innovation management are evolving: it is no longer about introducing innovations by an individual enterprise but about the transformation of innovations in an open model. Innovation has also ceased to be the result of a sequential process [8]. The new model has become a model that is comprised of one leader but is based on feedback from research activities of various companies. The researchers distinguish three models of innovation management, that is, a model of sustainable activities, network model, and a triple helix model [9–11].

a. The model of sustainable activities of the organisation was developed by the Japanese in the 1980s. In this model, various functional units of the company carried out simultaneous activities for the purpose of innovation development.

b. The network model, in which innovation is created through the process implemented by networked enterprises and the development of innovation is guaranteed by strategic alliances of research and development departments of various companies.

c. The triple helix model proposes the development of innovation between governmental organisations and universities [12–15].

The research showed that defining innovation management models differs fundamentally between individual companies and sectors. Automotive companies, which had so far generated innovations in their own functional units, now face new challenges. They are now trying to develop innovations in the open innovation model (either in the network model or the triple helix model).

The two-dimensional concept of measuring the effects of the innovation management model [16] introduces the concept of innovation performance (or innovation), defining it as: the quantity and quality of innovative ideas and the effectiveness and efficiency of implementing these ideas. The above-mentioned parameters of innovation performance can be examined independently and interdependently.

The analysis of innovation performance requires clarification of indicators (measuring the quantity, quality of ideas, and effectiveness and efficiency of innovative processes), selection of methods for their research, and identification of elements (variables) affecting the performance of innovation. This analytical effort leads to understanding and improving the functioning of the innovation management model.

To prepare for the analysis of innovation performance, a literature review was conducted in order to search for indicators, methods, and dependent and independent variables. According to the review [17–19], the innovativeness of companies is most often measured by expenditure on research and development and the number of registered patents. These are conventional indicators (still eagerly used due to their availability). Over time, these indicators have been criticised in the literature as having many disadvantages, and enriched or replaced by survey indicators (developed by the OECD and Eurostat). The quality of the proposed innovative solutions is measured by added value (economic, market, ecological, and social). The effectiveness of research and development processes is based on the number of implemented and commercialised product innovations and length of cycles of product innovation implementation. The effectiveness of processes is most often measured by the profits generated during the sale of innovation and the durability of innovation. If the researcher uses all of the mentioned indicators, they apply the so-called Inex. The applied innovation performance indicators of enterprises in the published studies are shown in **Figure 1**.

In the study of innovation performance, terminologies and methods adapted for this purpose by researchers [19–22] are used most often. A limited number of researchers focus on the study of the relationship between the company's innovation performance and the cooperating companies' innovation performance (For this purpose, statistical methods are used by few researchers). Authors using statistical methods most widely used multiple regression techniques, the least squares

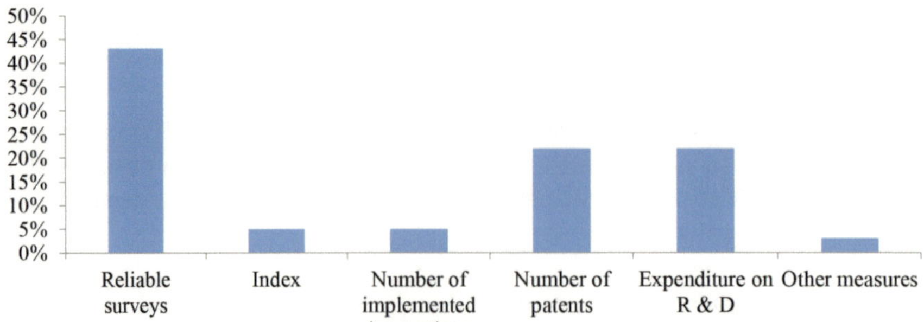

Figure 1.
Applied innovation performance indicators of enterprises [19].

method, logarithmic methods, and probit models (39.3% of the analysed studies used these methods). The researchers also used other statistical econometric methods (23.8%), such as factor analysis, cluster analysis, principal component analysis, count data models, and stochastic border analyses [23].

In 1998, the groundbreaking article "Research, Innovation, and Productivity: An Econometric Analysis at the Firm Level" by Crépon, Duguet, and Mairesse was published. In this article, the researchers created an empirical framework called CDM (the acronym of the three authors' names) or Mairesse Model. The framework is used in almost every study on innovation performance [24]. The CDM has inspired other researchers who over the years have proposed many variations of the original CDM model depending on whether they use continuous or discrete data to study innovation, different output measures, and different estimation methods. The original cross-sectional model has been expanded to include panel data, dynamic models, and applications for many types of innovative activities and indicators such as economic performance, that is, profitability or productivity of the enterprise.

As mentioned, the analysis of innovation performance cannot be deprived of a review of dependent, independent, and the so-called contextual/conditional variables. The literature lists those variables that affect the innovation performance of enterprises. Among the contextual variables, cooperation in the supply chain is listed first. Therefore, in the next chapters of the article, innovation performance will be presented in the context of using suppliers' resources from the supply chain.

3. Management of relations and resources of cooperating companies: the role of the integration platform in shaping the supply chain and increasing its innovation performance

The necessary factors for obtaining high innovation performance of a company integrating the supply chain are optimally selected resources and competences of individual participants in terms of quantity, quality, and cost, adequate flow of data, information, and knowledge, supported by appropriate IT tools [25, 26].

Standard registration systems or a database easily subjected to analytical processing are no longer sufficient. Business intelligence tools equipped with knowledge of resource theory, dominance, and network approach are needed, as well as tools testing different theories on the basis of many paradigms, drawing from the literature classifying management paradigms. On the basis of the network paradigm variables, six basic outsourcing models were built at the University of Tennessee, which dominate in the twenty-first century [27]. One of them is the vested outsourcing model, which postulates the transition from a traditional business model, oriented to a buy/sell transaction, to a result-oriented and value-added model, in which all business partners benefit. Business intelligence development work is based on this model-supporting innovative management in the supply chain.

Innovative websites and integration platforms based on the vested model are being introduced. One such platform is the Jabil Supply Chain Operations Center. The platform can simulate the supply chain from the first to the last link in real time. The system identifies high, medium, and low risk flows in the supply chain, as well as the possibilities of their optimisation. It provides supply and processing of data on the resources of individual enterprises belonging to the supply chain, labour law, taxes, admissibility of goods, political unrest, changing product requirements, breakdowns, and congestion. Jabil Supply Chain can assess all those factors that proactively affect the configuration of the supply chain, optimising it for each customer or for all at once. The chain can be modelled at the beginning of the product life cycle, that is, in the research and development phase (before commercialisation), and in the subsequent phases. The platform can also deepen the analyses of the logistics operations or simulate the shape of a new logistics chain. **Figure 2** presents the architecture of a business integration platform (identifying data sources for analysis, data collection, storage and processing systems, and information that can be generated).

Another example, also known as the new hub for innovation, is the Jabil Blue Sky Center. This platform brings together key experts from around the world (about 350 people) and connects businesses through these people. The world's largest curved touch screen serves the purpose of combining R&D activities, giving the possibility of illustrating a wide portfolio of engineering abilities of users logging in and at the same time stimulating the imagination of the creators of the idea of a new product.

An example of a company shaping its supply chain based on the Jabil Supply Chain solutions is Tesla Motors. Tesla is building a gigantic factory in which batteries for electric vehicles will be produced. Suppliers must participate in this unprecedented scale of production. The supply chain is currently shaped by the platform. Smaller companies specialising in battery technologies can join it. The database located on the platform receives online data from suppliers from around the world (in a predefined standard). For example, REDT (a company producing vanadium flow batteries) understands the importance of a contract with such a large partner so it provides the necessary data. It chose the Jabil database, thus entrusting this platform with confidential technological data and valuable production knowledge, hoping that they will join the Tesla Motors supply chain network [29].

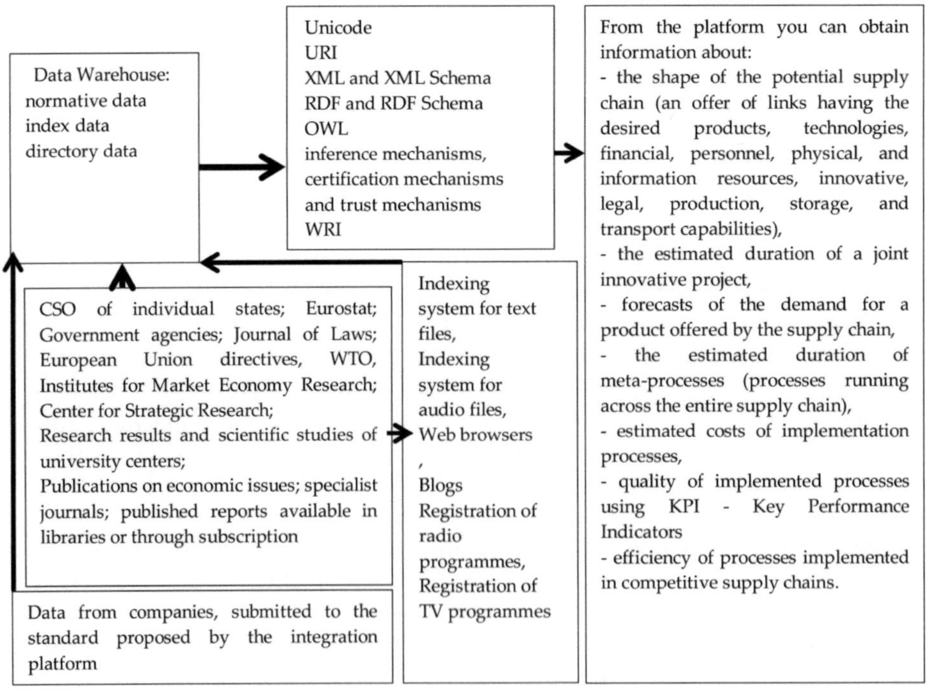

Figure 2.
Architecture of the system analysing the economic environment for the needs of shaping and managing the supply chain [28].

4. Advanced ICT tools for supporting knowledge management in order to increase the innovation performance of the supply chain

Automotive companies implement suppliers' innovation management. They employ managerial methods and tools that shape innovative supply chains and increase their innovation performance. The currently mentioned activities cannot be called systemic because the innovativeness of suppliers and their impact on the integrator's innovation performance is not measured [25–27, 30].

The Toyota company is the closest to a systemic innovation management of suppliers. Among the keiretsu group suppliers, Toyota builds a specific organisational culture and uses the philosophy of challenge, kaizen, genchi genbutsu, mutual respect, teamwork, Toyota Production System, and quality circles. The supplier innovation management system implemented by Toyota is a perfect example of the Japanese approach to teamwork. Decisions on developing products are not made suddenly: they are worked out during many meetings, discussions, and collective consultations in the group of managers of cooperating enterprises. Even if Toyota's actions are not explicitly referred to as a management system, they in fact create an integrated set of rules, procedures, and methods oriented at creating, popularising, and using knowledge about innovations. Toyota includes international suppliers in the development of process and product innovations.

Toyota uses a lot of ICT tools for creative problem-solving, for example, IWB (Innovation WorkBench). The software schematically presents problems, automatically analyses them, and leads users to an abstract solution; IAP (Innovation Assessment Program) is software invented by the United Inventors Association. The software helps inventors, entrepreneurs, and marketing specialists in conducting objective analyses of opportunities and threats for ideas and inventions [31]. Toyota develops innovative car software together with suppliers using the standard open source code. The Automotive Grade Linux (AGL) community in Linux Foundation has built an open source platform. Toyota used the platform to develop an audio and infotainment system launched on the market in 2018 and installed in Toyota Camry.

ICT tools have much to offer in the area of initiating innovations arising in the open model in the automotive industry. Research confirming this thesis focused on the phenomenon of open source free software [32], crowdsourcing platforms [33, 34], and network innovation brokers [35]. Other researchers analysed the contribution of ICT to the ability to absorb knowledge from the outside of the enterprise [36], data mining, simulation, prototyping, and visualisation supporting cooperating enterprises in developing new products [37].

Currently, the software for designing in an open model, based on the Blockchain & Smart Contract technology, is developing intensively. The Networking Innovation Room (NIR) is an innovative model of intellectual property (IP) protection created by a group of enterprises. Under NIR, the use of the Non-Disclosure Agreement (NDA) is proposed as an intelligent contract in which the remuneration is the so-called Wits, a virtual currency based on knowledge interactions [38]. Blockchain is a peer-to-peer platform that uses ICT to track the ownership of generated and transferred assets in an open innovation model [39]. Intelligent inter-organisational contracts are run and stored in Blockchain [40]. NIR controls the values added by suppliers, thus reducing the concerns of enterprises regarding the loss or underestimation of the intellectual property contribution. In the concept of NIR, companies in the SME sector [41] are particularly cared for. Everything that is submitted by companies is disclosed in NIR, timestamped, indexed, preserved, made searchable and traceable, and reported (when cooperating companies require it). An NDA is digitally accepted and can be signed when user enters the NIR. The agreement clearly describes the regime of intellectual property protection and the manner in which the co-created innovation will be protected [42]. The function of the developed application is signing legally binding, intelligent contracts that are produced using artificial intelligence creating a blockchain record path. This process is also called "IP document notarisation". Inventions, projects, and evidence can be quickly registered and the blockchain certificate will confirm the ownership, existence, and sustainability of the IP. All secured notarial information will remain private thanks to cryptography. Blockchain platforms, with the possible functional composition: PoE timestamp; integrity and notarial confirmations; IP register; content metadata; user authentication; record keeping; access control; licensing; traceability; quotation monitoring; reward mechanisms; own currency; NDA management; register of industrial property; and proof of receipt [40].

Some of the innovations that are being developed require fintech services [43]. It is possible to monitor the workflow throughout the innovative community. The system coordinates agreeing on what each player has to do in the cooperation process, when and what corrective actions would apply, what rewards or penalties are used for achieving/not completing milestones, tasks, etc.

CryptoTech also presents many other cases of the use of intelligent contract in the automotive sector's supply chains [44].

5. Research methodology–stages, objectives, hypotheses, research model

The first chapter lists the indicators of the innovation performance of the enterprise. Indicators evaluate but there are no such ones that allow the impact of the innovation management of supply chain companies on the innovation performance of the leader to be measured. There are no indicators measuring the impact of the integrator's influence on the chain links. The basic objective of the research was to verify the usefulness of the research tool empirically, which-following adjustments-could be used in the practice of the innovation performance management of automotive companies and their supply chains.

The second and third chapter looked at ICT methods and tools to help manage the innovation of suppliers. Methods and tools used by automotive companies were recognised and described [cedewu]. The previous studies by the author (carried out with the use of the diagnostic survey method) revealed that Toyota, Volkswagen, and FCA incur expenditure on the development of OI.

The expenditure incurred on the development of suppliers' innovativeness is not analysed in those studies but it should be monitored in order to optimise it. It is recommended to divide expenditure on the management of the supply into expenditure on the development of suppliers' competences (expenditures on development programmes) and expenditure on managerial tools using ICT technology. To make the management of supply chain innovation systemic, one should measure the amount of expenditure and draw practical conclusions for making subsequent investment decisions. Meanwhile, the automotive sector (in the automotive companies studied) does not measure suppliers' innovation or their impact on the integrator's innovation performance. The system that is proposed does not register its own expenses on the suppliers' development programme and ICT tools that support the suppliers' innovation management.

The presented research attempts to record expenditure on the development of suppliers' innovation. The objective of the research was to determine the correlation of innovation performance and the expenditure incurred on research and development activities, suppliers' development programmes. and ICT tools facilitating

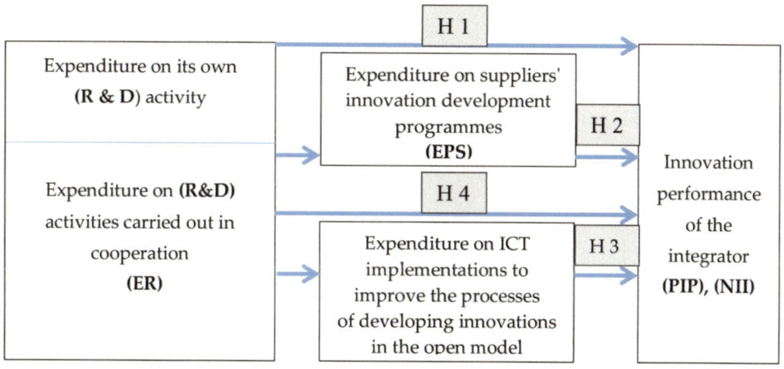

Figure 3.
Research model (source: author's own study).

innovative cooperation with them. The implementation of the research objective allowed for the empirical verification of the usefulness of the prepared research tool, which after the adjustments will be able to be implemented into the practice of managing the innovative.

The implementation of the objective required four hypotheses:

H1. Expenditure on its own research and development activities improves the innovation performance of the company.

H4. Expenditure on research and development in the open innovation model with suppliers influences the increase in innovation performance of the integrator. Increasing expenditure on suppliers' innovation development programmes results in an increase in the innovation performance of the integrator–H2. Increasing expenditure on ICT implementations (improving the processes of developing innovations in the open model) increases the innovation performance of the integrator–H3. The research model presented in **Figure 3** was prepared.

The determination of the relationship between the expenditure and innovation performance of automotive companies took place in the following research stages:

A. The following variables were operationalised:

- Expenditure on R&D activities *in EUR million* (published data).
- Expenditure on R&D activities carried out together with deliveries (ER) *in EUR million* (value estimated by respondents as a percentage of R&D expenditure).
- Expenditure on suppliers' innovation development programmes (EPS) *in EUR million* (value estimated by respondents).
- Expenditure on ICT implemented in order to improve processes of innovation development with suppliers (ECT) *in EUR million* (value estimated by respondents as a percentage of expenditure on CapEx).
- Expenditure on R&D activities for net sales per year (NII) *in %* (published data-Intel R&D).
- Profits generated on innovative products (PIP) *in EUR million* (value estimated by respondents as a percentage of operating profit).

B. A compilation of published data was prepared, which is presented in **Table 1**.

World rank	Company	Country	Expenditure on R&D (€million)	R&D one-year growth (%)	Net sales (€million)	Net sales one-year growth (%)	R&D intensity (%)	CapEx (€million)	CapEx one-year growth (%)	CapEx intensity (%)	Op. profits (€million)	
			2016/2017									
1	VW	Germany	13672.0	0.4	217267.0	1.9	6.3	13152.0	−0.5	6.1	8344.0	3.8
13	TOYOTA	Japan	7500.1	−12.5	224150.8	−2.8	3.3	28764.4	−12.8	12.8	16198.7	7.2
34	FCA	Netherlands	4219.0	2.7	111018.0	0.4	3.8	8815.0	0.0	7.9	5109.0	4.6
			2015/2016									
1	VW	Germany	13612.0	3.8	213292.0	5.4	6.4	13213.0	10.0	6.2	−1228.0	−0.6
10	TOYOTA	Japan	8047.0	5.1	216506.5	4.3	3.7	30941.9	20.9	14.3	21754.8	10.0
31	FCA	Italy	4108.0	12.1	110595.0	15.1	3.7	8819.0	8.6	8.0	2625.0	2.4
			2014/2015									
1	VW	Germany	13120.0	11.7	202458.0	2.8	6.5	12012.0	5.5	5.9	12139.0	6.0
9	TOYOTA	Japan	6858.4	10.3	185940.4	6.0	3.7	22923.4	25.3	12.3	18779.1	10.1
30	FCA	Netherlands	3665.0	9.0	96090.0	10.9	3.8	8121.0	8.4	8.5	3343.0	3.5
			2013/2014									
1	VW	Germany	11743.0	23.4	197007.0	6.0	15.8	11385.0	8.5	5.8	11500.0	5.8
7	TOYOTA	Japan	6269.9	12.8	177017.3	3.5	10.6	18456.2	35.7	10.4	15792.7	8.9
32	FIAT	Italy	3362.0	2.0	86816.0	3.9	34.3	7440.0		8.6	3499.0	4.0
			2012/2013									
1	VW	Germany	9515.0	32.1	193000.0	4.9	22.5	10493.0	29.8	5.4	8333.0	4.3
5	TOYOTA	Japan	7070.9	3.5	193000.0	3.7	5.2	17287.7	28.9	9.0	11567.0	6.0
34	FIAT	Italy	3295.0	51.5	83957.0	3.9	18.8	7534.0	36.3	9.0	3921.0	4.7

World rank	Company	Country	Expenditure on R&D (€million)	R&D one-year growth (%)	Net sales (€million)	Net sales one-year growth (%)	R&D intensity (%)	CapEx (€million)	CapEx one-year growth (%)	CapEx intensity (%)	Op. profits (€million)	
			2011/2012									
1	Toyota	Japan	7754.5	7.6	184,798,1	−1.9	4.2	15235.2	6.6	8.2	3536.4	1.9
3	VW	Germany	7203.0	15.1	159337.0	25.6	4.5	8087.0	−36.7	5.1	10930.0	6.9
52	FIAT	Italy	2175.0	12.3	59559.0	4.1	3.7	5528.0	112.3	9.3	3442.0	5.8

Table 1.
Published data related to R&D activities performed by the surveyed companies [45].

C. A questionnaire was prepared. The questionnaire is included in **Annex 1**. The questionnaire was sent electronically, using public responder portals of automotive companies. The necessary data were obtained (with the indication that these are only calculated estimates, not supported by a thorough analysis of the finances of the surveyed companies).

D. The obtained data were processed statistically. Calculated correlations and regressions confirmed the hypotheses H1, H2, H3, and H4. Summary and conclusions drawn from the canonical analysis are the culmination of research and confirmation of hypotheses.

The correlation analysis looked primarily at the relationship between (ER), (EPS), (ECT), and innovation performance of automotive companies (NII) and (PIP).

Expenditure on research and development (implemented in the open model with suppliers) (ER), expenditure on suppliers' innovation development programmes (EPS), and expenditure on ICT implemented to improve the processes of obtaining innovations (ECT) were treated as dependent variables. The ratio of expenditure on R&D to net sales (NII) and profits generated on innovative products (PIP) were treated as independent variables. Literature theses were used to confirm the dependence of these variables. The operationalisation of profits from implementations of innovative product solutions was revised [46]. Jałowiecki dealt with the impact of implementing ICT on profits from innovative products [47]. On the other hand, Dzikowski [48] believed that the amount of profits from innovative products depends on the expenditure on suppliers' innovation development programmes. Brem and Tidd [49] believed that the use of a research model as a simulation model supporting innovation management could lead to improved results (independent variables). The use of correlation and regression analysis (on historical data) in innovation management facilitates making investment decisions.

6. Comparative analysis of the innovation performance of Toyota, Volkswagen, and FCA using the research model developed

Figure 4 shows the intensity of R&D in percent, that is, the ratio of expenditure on research and development (in EUR million) to net sales (in EUR million) in the surveyed companies.

Figure 5 shows the intensity of CapEx in percent, that is, the ratio of investment expenditure (in EUR million) to fixed assets (including ICT software) for net sales (in EUR million) in the surveyed companies.

Figure 6 shows the intensity of R&D activities in OI with suppliers in percent (or ER intensity), that is, the ratio of expenditure on research and development made with suppliers (in EUR million) to net sales (in EUR million) in the companies surveyed.

Figure 7 shows the intensity of expenditure on IT systems used to develop innovations in the open model in percent (or ECT intensity), that is, the ratio of IT expenditure (in EUR million) to net sales (in EUR million) in the companies surveyed.

It can be seen from the figures that Volkswagen bears the largest expenditure on research and development in relation to sales, but it invests in ICT (**Figure 7**) with an emphasis on expenditure in the closed innovation model (as shown in **Figure 6**). Toyota, on the other hand, invests in production systems and ICT (as shown in **Figures 5** and 7) and more than other companies invest in research and development in the open model with suppliers (as shown in **Figure 6**). FCA implements expenditure on R&D and invests in systems in a sustainable manner (as shown in

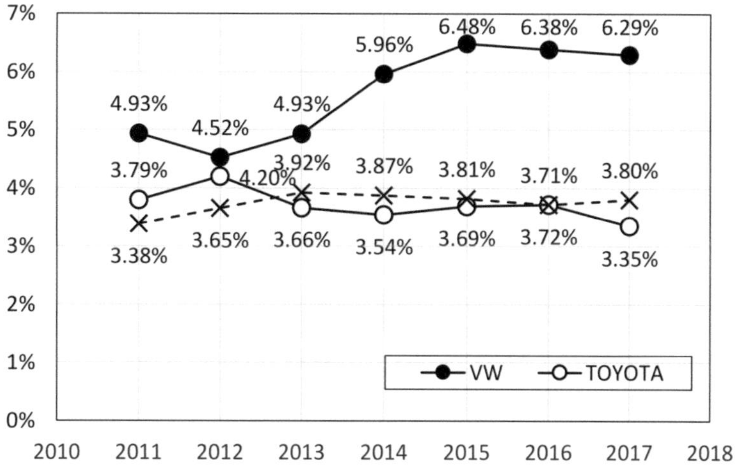

Figure 4.
R&D intensity (%) in the companies surveyed [45].

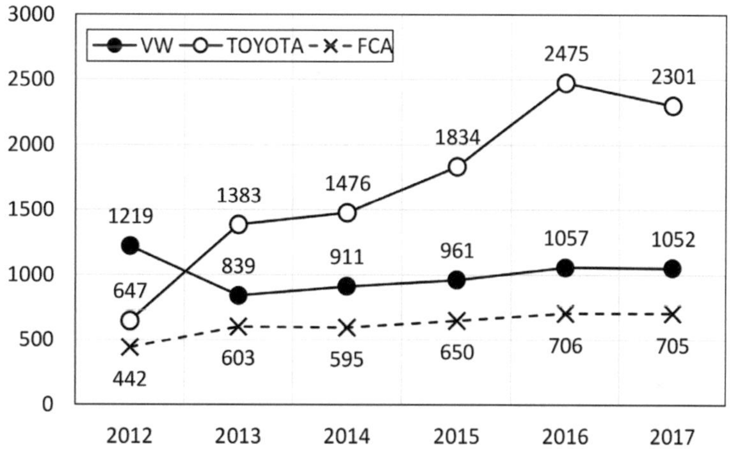

Figure 5.
CapEx intensity (%) in the companies surveyed [45].

Figures 4 and **5**). It maintains a constant trend in incurring expenditure on R&D in the open model (as shown in **Figure 6**).

There are interesting conclusions that can be drawn based on a comparison of changes of investment intensity in research and development over time (**Figure 4**) and investing in research and development but implemented in cooperation with suppliers (**Figure 6**) in Toyota. **Figure 4** clearly shows that the data for this company have a 4-degree polynomial distribution. This is indicated by the high value of the R^2 determination coefficient, amounting to 0.9743 of the theoretical model to empirical data. Such a distribution is rarely seen in practice, with one exception. If the data are a time series and the theoretical model is adapted to its fragment of the appropriate length, it is likely that we are dealing with a time series devoid of or characterised by a negligible trend and clearly marked periodic fluctuations. In contrast to seasonal variations, its amplitude is not 1 year but an appropriate number of periods. In case of changes in the intensity of investing in research and development presented in **Figure 4**, seasonal variations of 5 years are most likely. Of course, the data from the previous years should be checked. Nevertheless, the

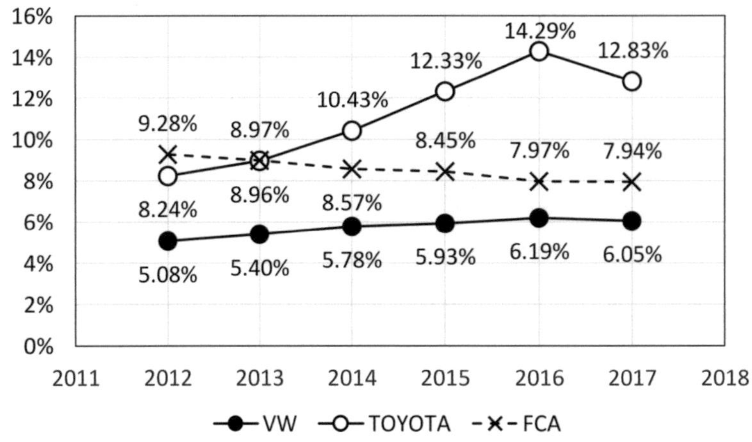

Figure 6.
Intensity of R&D in OI (%)/ER intensity (%) in the companies surveyed (source: author's own study).

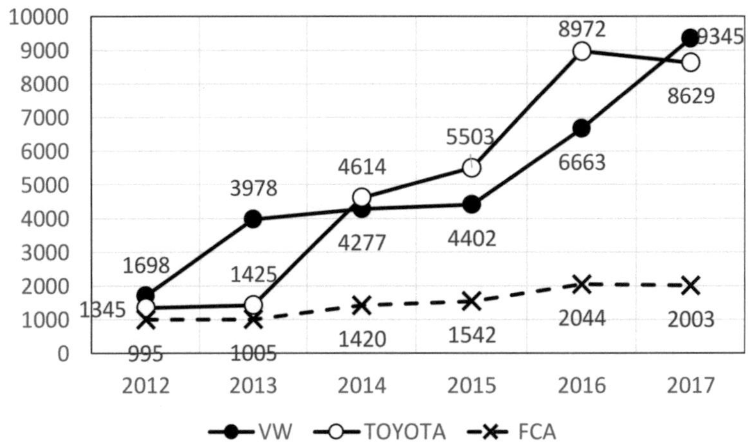

Figure 7.
Intensity of ECT (%) in the companies surveyed (source: author's own study).

stability of the time series with only periodic fluctuations is clearly visible. Toyota is a company with an established position in the market (which also applies to the entire Japanese economy); therefore, research and development expenditure is stable. Meanwhile, **Figure 5** clearly shows that the time series for investment in research and development in Toyota's cooperation with suppliers shows a clear upward trend. This means that the volume of this characteristic group of investments in research and development is systematically increasing. According to the Tapscott concept, this is a characteristic feature of all highly developed digital economies, which are called networks. In such economic systems, the level of investments in cooperative research and development increases significantly.

The following **Tables 2–5** show the correlation of operational variables for Toyota, VW, and FCA respectively, and the analysis of correlations from variables being the sum of data from the three surveyed companies.

The correlation analyses presented above reveal that in Toyota the variable (PIP) reflecting innovation performance and the variables (ER) and (ECT) achieved high correlation rates, which can be interpreted as follows: Toyota's innovation depends on cooperation with suppliers, and this cooperation largely depends on ITC systems connecting cooperating innovators.

Variables	(B + R)	(ER)	(EPS)	(ECT)	(NII)	(PIP)
(R&D)	0	—	—	—	—	—
(ER)	0.779	0	—	—	—	—
(EPS)	0.505	0.779	0	—	—	—
(ECT)	0.651	0.873	0.684	0		
(NII)	0.577	0.643	0.959	0.846	0	
(PIP)	0.430	0.630	0.899	0.866	0.974	0

Table 2.
Analyses of correlation of variables related to Toyota's R&D activity (source: author's own study).

Variables	(B + R)	(ER)	(EPS)	(ECT)	(NII)	(PIP)
(R&D)	0	—	—	—	—	—
(ER)	0.884	0	—	—	—	—
(EPS)	0.759	0.573	0	—	—	—
(ECT)	0.889	0.974	0.633	0		
(NII)	0.761	0.652	0.977	0.726	0	
(PIP)	0.879	0.767	0.777	0.843	0.834	0

Table 3.
Analyses of correlation of variables related to Volkswagen's R&D activity (source: author's own study).

Variables	(B + R)	(ER)	(EPS)	(ECT)	(NII)	(PIP)
(R&D)	0	—	—	—	—	—
(ER)	0.978	0	—	—	—	—
(EPS)	0.951	0.978	0	—	—	—
(ECT)	0.918	0.921	0.966	0		
(NII)	0.538	0.446	0.756	0.708	0	
(PIP)	0.740	0.724	0.849	0.812	0.807	0

Table 4.
Analyses of correlation of variables related to FCA's R&D activity (source: author's own study).

Variables	(B + R)	(ER)	(EPS)	(ECT)	(NII)	(PIP)
(R&D)	0	—	—	—	—	—
(ER)	0.948	0	—	—	—	—
(EPS)	0.818	0.948	0	—	—	—
(ECT)	0.987	0.980	0.768	0		
(NII)	0.794	0.637	0.995	0.748	0	
(PIP)	0.931	0.784	0.952	0.886	0.928	0

Table 5.
Analyses of correlation of variables related to R&D activity-sum of T, V, F (source: author's own study).

In addition, the above correlation analyses show that in Volkswagen the variable (PIP) reflecting innovation performance and the variables (R&D) and (ECT) achieved high correlation rates, which can be interpreted that VW's innovation depends primarily on its own R&D departments and expenditure on IT systems.

The above correlation analyses show that in FCA the variable (PIP) illustrating innovation performance and other variables obtained similarly high correlation rates. This means that FCA manages its innovation performance in a sustainable way.

Table 6 presents the analysis of correlations from variables being the sum of data from the three surveyed companies. A linear regression analysis was also carried out on the basis of aggregated data, which allows the simulation to be performed. It revealed that the following:

- An increase of 1% in expenditure on suppliers' development programmes (EPS) results in an increase in profits from the sale of innovative products (PIP) by 1.87% on average, as shown in **Figure 8**.

- The increase in ICT expenditure implemented to improve the efficiency of innovation processes with suppliers (ECT) by 1% results in an increase in profits from the sale of innovative products (PIP) by 0.18% on average, as shown in **Figure 9**.

Similar simulations will be the subject of further research, and decision-makers will be able to optimise expenditure on research and development activities implemented within the company and in cooperation.

Variable	Year							
	2010	2011	2012	2013	2014	2015	2016	2017
R&D*								
ER								
EPS								
ECT								
NII*								
PIP								

**http://iri.jrc.ec.europa.eu/scoreboard.html*

Table6.
Table for collecting data for correlation analysis in the "Model of innovation performance research of the supply chain leader".

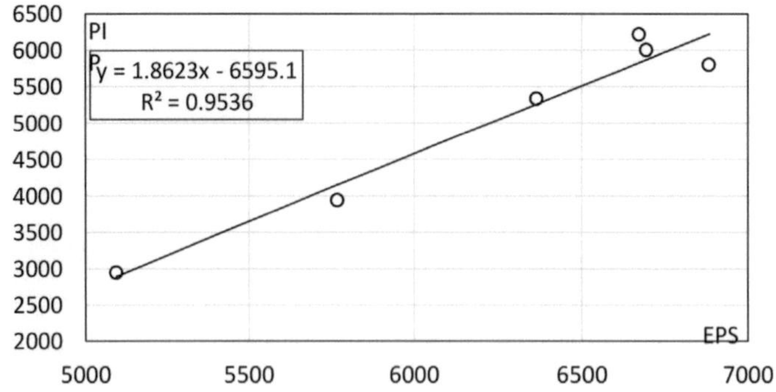

Figure 8.
Expenditure on suppliers' development programmes and profits from the sale of innovative products (source: author's own study).

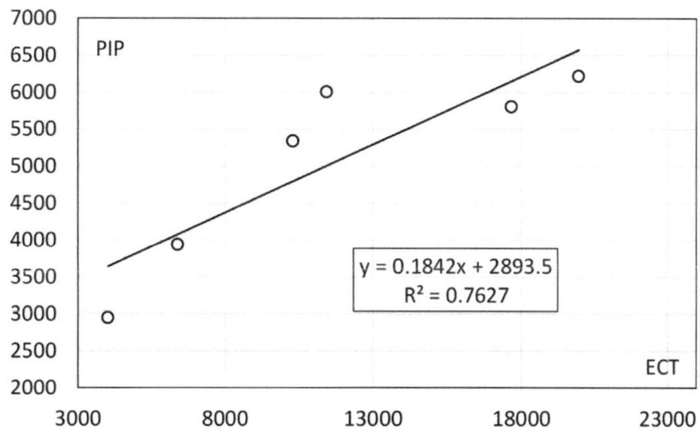

Figure 9.
Expenditure on ICT for OI and profits from the sale of innovative products (source: author's own study).

7. Discussion of the research results

Attempts to obtain quantitative data have shown that companies do not accumulate them in a way that would clearly identify expenditure on R&D in the open and closed model. Expenditure on suppliers' innovation development programmes is not measured. In addition, it was difficult for the companies to select expenditure on ITC (implemented for supporting research and development processes carried out in cooperation with suppliers). The analyses performed above further confirmed that the companies use simple indicators of their own and suppliers' innovation performance (although researchers are already offering more advanced assessment methods).

The researcher notices a number of limitations for the conducted research, affecting the reliability of the presented analyses. All correlation analyses should be based on reliable, publicly available financial documents. The correlation indicators presented in the paper may, however, contain errors, as their calculations are based on estimated data. The estimated data were received from the automotive companies' central offices (through responders-response portals of surveyed companies, using the questionnaire—**Annex 1**) as a percentage of the values reported in official documents. It was argued that corporations do not keep records of expenditure in the layout desired by the created research model. The implementation of the suppliers' innovation management model means defining and monitoring indicators that make it possible to measure expenditure on research and development in the open model in the future.

8. Conclusions

Estimated analyses clearly show that the increase in expenditure on research and development activities in the open innovation model translates into an increase in the innovation performance of the supply chain integrator. It was shown that optimising expenditure on research and development activities (i.e., gradual abandonment of expenditure on internal R&D activities and an increase in expenditure on the development of suppliers' innovation) results in a significant increase in the innovation performance of automotive companies.

It was considered right to prepare a digital model for suppliers' innovation management that would be implemented in an automotive company. Its primary goal would be to optimise expenditures on research and development activities carried out together with suppliers. While preparing the model, the leading role of the company in the supply chain and its irreplaceable influence on improvements to suppliers' innovation were recognised. It seems that the surveyed companies still have an indifferent attitude towards new ICT technologies used to develop innovations in the open model. They invest little in modern ICT systems and do not measure the effectiveness of these expenditures.

Summarising the attempt to implement a digital model of suppliers' innovation management, it should be noted that an analytical look at the R&D field is not possible right now due to the lack of properly collected data. The proposed model of suppliers' innovation management could be implemented in the IT system of the automotive company and serve as a tool supporting the development of innovation in the open model. The idea of implementation was discussed with a specialist-an employee of the SAP company (who is responsible for cooperation with companies from the automotive industry). The idea was found to be of interest and confirmed feasible.

The presented research will further deepen the discussion on digital innovation management of suppliers in the supply chain.

Acknowledgements

The translation of the article was financed with BK-235/ROZ0/2018 implemented by the Silesian University of Technology, ul. Akademicka 2A, 44-100 Gliwice.

Conflicts of interest

The author declares no conflict of interest.

A. Annex 1

A.1 Suppliers' innovation management

Dear Sir or Madam. Correlation of variables from this table is the culmination of my research work. I would like to show that including suppliers in the process of developing innovative products influences the profits of automotive companies. I am asking you for approximate percentages only. Thank you in advance for these data/variables:

R&D Expenditure on R&D.

ER R&D expenditure realised jointly with suppliers *in EUR million* (estimated by the respondent as a percentage of expenditure on R&D).

EPS Expenditure on suppliers' innovation development programmes in EUR millions (estimated value by the respondent, companies do not distinguish these costs).

ECT Expenditure on ICT to improve the processes of innovation development with suppliers in EUR millions (estimated by the respondent as a percentage of expenditure on CapEx).

NII Expenditure on R&D per year/net sales per year-Intensive R&D).

PIP Profits generated on innovative products (as a percentage of operating profit).

Glossary of used terms and abbreviations

R&D	(Research and Development); usually team activities of scientific or technical nature. Activities can be divided into basic, applied, and developmental research.
OI	(Open Innovation)-A concept popularised by the professor and executive director of the Open Innovation Center at the University of Berkeley, Henry Chesbrough. According to this concept, enterprises should not rely solely on the results of their own research and development work but use external sources of innovation through cooperation with other entities.
ICT	(Information and Communication Technologies)-Information and communication technologies. The concept includes technologies that process, collect, and send information in electronic form.
NDA	(Non-Disclosure Agreement)-a contract of confidentiality that obliges its parties to exchange confidential materials or information subject to their further non-dissemination. The NDA is also known as the CDA (confidential disclosure agreement).
CDM	(acronym of the three authors' names) or Mairesse Model-an econometric analysis used at the company level to study innovation and productivity, introducing concepts and definitions: IP-Innovation Performance/Innovation.
IP	Innovation Performance/Innovation. The assessment of IP consists of: profits from the sale of innovative products, the number of implementations of innovative solutions, the quality of implemented innovative solutions, effectiveness and efficiency of product development processes.

Author details

Ewa Stawiarska
Silesian University of Technology, Poland

*Address all correspondence to: ewa.stawiarska@polsl.pl

References

[1] Chapman RL, Soosay C, Kandampully J. Innovation in logistic service and the new business model: A conceptual framework. Managing Service Quality. 2002;**12**(6):358-371

[2] Brem A, Voigt K-I. Integration of market pull and technology push in the corporate front end and innovation management—Insights from the German software industry. Technovation. 2009;**29**:351-367

[3] Koen P, Ajamian G, Burkart R, Clamen A, Davidson J, D'Amore R, et al. Providing clarity and common language to the "fuzzy front end". Research-Technology Management. 2001;**44**(2): 46-55

[4] Cooper RG. How companies are reinventing their idea-to-launch methodologies. Research-Technology Management. 2009;**52**(2):47-57

[5] Herstatt C, Verworn B, Nagahira A. Reducing project related uncertainty in the "Fuzzy Front End" of innovation–A comparison of German and Japanese product innovation projects. In: Herstatt C, Stockstrom C, Tschirky H, Nagahira A, editors. Management of Technology and Innovation in Japan. Berlin: Springer; 2006. pp. 329-352

[6] Bocken NMP, Farracho M, Bosworth R, Kemp R. The front-end of eco-innovation for eco-innovative small and medium sized companies. Journal of Engineering and Technology Management. 2014;**31**:43-57

[7] Sheu DD, Lee H-K. A proposed process for systematic innovation. International Journal of Production Research. 2011;**49**(3):847-868

[8] Chesbrough H, Crowther AK. Beyond high tech: Early Adopters of open innovation in other industries. R&D Management. 2006;**36**(3):229-236

[9] Rothwell R. Industrial innovation: Success, strategy, trends. In: Dodgson M, Rothwell R, editors. The Handbook of Industrial Innovation. Hants: Edward Elgar; 1994

[10] Etzkowitz H, Leydesdorff L. The dynamics of innovation: From national systems and "mode 2" to a triple helix of university-industry-government relations. Research Policy. 2000;**29**(2): 109-123

[11] Stuart TE. Interorganizational alliances and the performance of firms: A study of growth and innovation rates in a high-technology industry. Strategic Management Journal. 2000;**21**(8): 791-811

[12] West J, Gallagher S. Challenges of open innovation: the paradox of firm investment in open-source software. R&D Management. **36**(3):319-331

[13] Chesbrough HW, Appleyard MM. Open innovation and strategy. California Management Review. 2007; **50**(1):57-76

[14] Laursen K, Salter AJ. Open for innovation: The role of openness in explaining innovation performance among UK manufacturing firms. Strategic Management Journal. 2006;**27**:131-150

[15] Dahlander L, Gann D. How open is innovation? Research Policy. 2010; **39**(6):699-709

[16] Ryan A. Innovation Performance. 2010. Available from: http://www.managedinnovation.com/articles [Accessed: 19 February 2018]

[17] Linton JD, Thongpapanl NT. Perspective: Ranking the technology innovation management journals. Journal of Product Innovation Management. 2004;**21**(2):123-139

[18] Thongpapanl NT. The changing landscape of technology and innovation management: An updated ranking of journals in the field. Technovation. 2012;**32**(5):257-271

[19] Becheikh N, Landry R, Amara N. Lessons from innovation empirical studies in the manufacturing sector: A systematic review of the literature from 1993–2003. Technovation. 2006;**26**(5): 644-664

[20] Schneider S, Spieth P. Business model innovation: Towards an integrated future research agenda. International Journal of Innovation Management. 2013;**17**(01): 1340001-134000134

[21] Crossan MM, Apaydin MA. Multi-dimensional framework of organizational innovation: A systematic review of the literature. Journal of Management Studies. 2010;**47**(6): 1154-1191

[22] Ozman M. Inter-firm networks and innovation: A survey of literature. Economics of Innovation and New Technology. 2009;**18**(1):39-67

[23] Lokshin B, Hagedoorn J, Letterie W. The bumpy road of technology partnerships: Understanding causes and consequences of partnership mal-functioning. Research Policy. 2011; **40**(2):297-308

[24] Crepon B, Duguet E, Mairesse J. 1998. Research, Innovation, and Productivity: An Econometric Analysis at the Firm Level, NBER Working Paper No. 6696. Available from: http://www.nber.org/papers/w6696 [Accessed: 19 February 2018]

[25] Stawiarska E, Dzikowski P, Bartczak K. Innowacyjność polskich przedsiębiorstw. Warszawa: Texter; 2017. pp. 13-45

[26] Stawiarska, E. The process of obtaining innovative solutions from suppliers and analysis of risks occurring in this process on the example of the automotive industry, Zeszyty Naukowe Politechniki Śląskiej, z. 2018;**119**:78-96

[27] Stawiarska E. Modele zarządzania innowacjami w łańcuchach i sieciach dostaw międzynarodowych koncernów motoryzacyjnych. Warszawa: CeDeWu; 2019. pp. 34-78

[28] Wincewicz-Bosy M, Stawiarska E, Łupicka A. Współczesne wyzwania łańcuchów dostaw. Text. 2017;**2017**: 113-126

[29] www.jabil.com. Available from: http://www.jabil.com/technologies/control-tower/ [Accessed: 14 February 2018]

[30] Stawiarska E. Managing green logistiscs chains for the automotive industry. In: 18th International Multidisciplinary Scientific Geo Conference. SGEM 2018, 2 July–8 July 2018, Albena, Bulgeria, Conference proceedings. Vol. 18. 2018. pp. 601-609

[31] Sorli M, Stokic D. Innovating in Product/Process Development. Vol. 140. London: Springer; 2009

[32] Hippel E, von Krogh G. Open source software and the "private-collective" innovation model: Issues for organization science. Organization Science. 2003;**14**(2):209-213

[33] Di Gangi, Wasko M. Steal my idea! Organizational adoption of user innovations from a user innovation community: A case study of Dell idea storm Decision Support Systems. 2009; **48**(1):303-312

[34] Leimeister JM, Huber M, Bretschneider U, Krcmar H. Leveraging Crowdsourcing: Activation-Supporting Components for IT-Based Ideas Competition. Journal of Management Information Systems. 2009;**26**(1): 197-224

[35] Whelan E, Golden W, Donnellan B. Digitizing the social network R & D; wracając do straży technologicznej. Information Systems Journal. 2013; **23**(3):197-218

[36] Chatterjee D, Pacini C, Sambamurthy V. The shareholder-wealth and trading-volume effects of information-technology infrastructure investments. Journal of Management Information Systems. 2002;**19**(2):7-42

[37] Dodgson M, Gann D, Salter A. The role of technology in the transition to open innovation: Coincidence Procter & Gamble. R&D Management. 2006; **36**(3):333-346

[38] Carrillo C, de la Rosa JL, Canals A. Towards a knowledge economy. International Journal of Community Currency Research. 2007;**11**:84-97

[39] Bogers M, Zobel AK, Afuah A, Almirall A, Brunswicker S et al. The Open Innovation Research Landscape: Established Perspectives and Emerging Themes Across Different Levels of Analysis, Forthcoming in Industry and Innovation. 2017. Available from: http://ssrn.com/abstract=2817865 [Accessed: 01 February 2018]

[40] Tapscott D, Tapscott A. Blockchain Revolution: How the Technology Behind Bitcoin is Changing Money, Business, and the World. Penguin Random House; 2016

[41] Bikfalvi A, de la Rosa JL, van Haelst S, Gorini M, Pelizzaro A, Haugk S. Study Report to Characterize the Target Groups in Relation to the Project Topics: SMEs and Innovation Advisors. Internal Report Published. 2015. Available from: http://dugi-doc.udg.edu//handle/10256/13269 [Accessed: 01 May 2018]

[42] Lusch RF, Nambisan S. Service innovation: A service-dominant logic perspective. MIS Quarterly. 2015;**39**(1): 155-175

[43] Kahan A. Legal Protection: Liability and Immunity Arrangements of Central Banks and..., IMF Working Paper, WP 18/176. 2017

[44] Blockchain for the Automotive Industry. Available from: https://www.cryptotec.com/wp-content/uploads/2018/01/Blockchain_for_Automotive_CryptoTec_EN.pdf [Accessed: 01 November 2018]

[45] Available from: http://iri.jrc.ec.europa.eu/scoreboard.html [Accessed: 01 May 2018]

[46] Se-Hwa W, Hsiang WH. Gatekeeping Mechanisms of Creative and Innovative Products. 2008. Available from: https://www.researchgate.net/profile/Se_Hwa_Wu [Accessed: 17 November 2018]

[47] Jałowiecki P. Paradoks produktywności Solowa w polskim przemyśle spożywczym. (Monograph), Warszawa: Wydawnictwo SGGW. 2018: 56-76

[48] Dzikowski P. Supply networks and innovation activity in medium-high technology manufacturing industries in Poland. Acta Scientiarum Polonorum Oeconomia. 2018;**17**(1):13-22

[49] Brem A, Tidd J. Perspectives on supplier innovation: Theories, concepts and empirical insights on open innovation and the integration of suppliers. Series on Technology Management Journal of Business-to-Business Marketing. 2014;**18**:57-62